IEE Electrical Measurement Series 4
Series editors Dr. A. C. Lynch
A. E. Bailey

The Current Comparator

Other volumes in this series:

The Current Comparator

W.J.M.Moore & P.N.Miljanic

Peter Peregrinus Ltd on behalf of The Institution of Electrical Engineers

Published by: Peter Peregrinus Ltd., London, United Kingdom

Moore, W. J. M.
    The current comparator—(IEE electrical measurement series; 4).
    1. Electronic transformers
    I. Title   II. Miljanic, P. N.
    III. Institution of Electrical Engineers
    IV. Series
    621.3815'1   TK7872.T7

ISBN 0 86341 112 6

Printed in England by Short Run Press Ltd., Exeter

Contents

Preface

Although three-winding transformers and transformer bridges have been in use for many years, the current comparator, as it is presently known and described in this book, was conceived at almost the same time by I. Obradovic, P. N. Miljanic and S. Spiridonovic of the Institut Nikola Tesla, Belgrade, Yugoslavia, and by N. L. Kusters and W. J. M. Moore of the National Research Council, Ottawa, Canada, in research directed towards the improvement of current transformer calibrations. In 1960, P. N. Miljanic joined the National Research Council where such advances as magnetic shielding, the compensation winding, and extensions to direct current soon followed. This work eventually led to current transformer calibration equipment of very high accuracy and stability which is easy to use and is relatively unaffected by ambient conditions.

The application of the current comparator in a bridge for high-voltage capacitance measurements and voltage transformer calibration was undertaken by N. L. Kusters and O. Petersons. This bridge was later adapted to the measurement of losses in inductive reactors and power transformers by W. J. M. Moore and E. So. O. Petersons continued his work at the National Bureau of Standards in Washington where he developed an improved bridge.

Low-voltage current comparator bridges for alternating current resistance, impedance and power measurements were developed by W. J. M. Moore, E. So and K. Ayukawa.

Following a feasibility study by N. L. Kusters, W. J. M. Moore and P. N. Miljanic, the direct current development was continued by N. L. Kusters and M. P. MacMartin. This led to a resistance bridge, a seven-decade potentiometer, and equipment for measuring large direct current ratios. Later E. So developed a 20-bit standard digital-to-analogue converter using direct current comparator techniques.

The success of this development could not, of course, have been achieved without the technical expertise and intuition of many others in the construction and evaluation of the various instruments. These include over the years (in alphabetical order) B. R. Cassidy, L. Chapman, M. J. Davis, P. Hawton, N. A. Lackey, G. Landry, L. MacNamara, P. Mougeot, E. Mulligan and L. Sabourin.

Of invaluable assistance also were Guildline Instruments Limited of Smiths Falls, Ontario, Canada and the Institut Nikola Tesla, Belgrade, Yugoslavia which, by producing versions of many of the instruments, made them available to laboratories and industry throughout the world.

Acknowledgement

The authors are indebted to Leslie A. Cameron for her illustrative artwork and to Doreen Dassen, Alyson Smith and Patti Lamoureux for their co-operation and patience in typing the several drafts of the manuscript.

Chapter 1

Introduction

1.1 Historical background

In the early 1950s, the electrical standards community was reawakened to the increased accuracy and stability that could be obtained in ratio measurements using transformers constructed with the new high-permeability, low-loss magnetic materials. In Australia, improved transformer-based voltage ratio bridges were being developed at the National Standards Laboratory (NSL) to exploit their newly discovered calculable standard of capacitance,[1] and these were soon followed by the National Bureau of Standards (NBS) in the United States.[2] In Germany, a set of four high-quality instrument transformers had been produced which provided the basis for an intercomparison of calibrations between the Physikalisch-Technische Bundesanstalt (PTB), the National Research Council of Canada (NRC), and NBS.

This intercomparison revealed to NRC that the accuracy of its calibration techniques was not good enough for those instrument transformers currently being manufactured, and a programme of improvement was therefore initiated. Studies of the three-winding current transformer, in which two of the windings carried the currents whose ratio was to be measured and the third was used to detect ampere-turn balance, indicated that stable and repeatable balances between the turns ratio and the current ratio were attainable to a few parts per 100 million.

The three-winding current ratio transformer is of course the conjugate of the three-winding voltage ratio transformer, the principal difference being that the power source and the balance detector are interchanged. There is however another more important difference. The current ratio transformer operates with zero flux in its magnetic core at balance and this, together with appropriate flux detection techniques, makes possible its application in direct current measurements.

1.2 The evolution of the alternating current comparator

An early application of the current comparator was in apparatus developed at the Institut Nikola Tesla, Belgrade, Yugoslavia for the testing of current transformers.[3] This comparator had primary and secondary ratio windings, a deviation winding, and a detection winding. Electronic circuitry, controlled by the detection winding, reduced the flux in the magnetic core to zero by supplying the error current to the deviation winding. Decomposition of this error current into in-phase and quadrature components was accomplished with two electrodynamic wattmeters whose potential coils were energised from an auxiliary electronically aided current transformer.

The first current comparators at NRC were constructed using magnetic toroidal cores wound from either 0·025 or 0·10 mm thick tapes of Supermalloy (initial permeability 41 000) or HyMu 80 (initial permeability 12 000). The detection windings were at first wound uniformly with a single conductor in two layers, one in reverse pitch to the other so as to avoid the formation of a loop or 'air turn' (Fig. 1.1a). This second layer however was soon abandoned in most designs, with the air turn being eliminated by bringing the other end of the winding back directly around the core (Fig. 1.1b) or with only a few turns of reverse pitch. This winding was then enclosed in a thick (≈0·5 mm) copper shield box with a gap along one edge to avoid forming a short-circuit around the magnetic core. The current ratio windings were formed by cabling several conductors together and winding them in one layer of about 30 turns once around the toroid. For a 1/1 ratio, half of the conductors were paralleled to form the primary winding and the other half were paralleled to form the secondary winding. In a 2/1 ratio version the secondary group of conductors was further divided into two, and these were connected in series to form a secondary

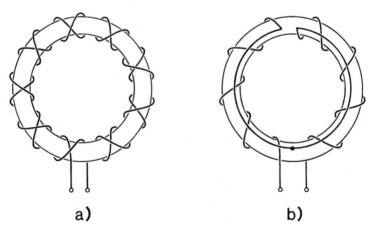

a)   b)

Fig. 1.1   *Detection winding layout. (a) Early arrangement, since abandoned (b) Present version with less shunt capacitance*

Fig. 1.2 *The first 5/5 and 10/5 ampere current comparators*

winding with double the number of turns and traversing twice around the toroid. The finished comparators were then enclosed in a copper box (Fig. 1.2). Their secondary current rating was 5 amperes.

These early comparators were used to establish methods for self-calibration at 1/1 ratio and to investigate step-up techniques for transferring these calibrations to higher ratios. Some important results of this work were as follows:

1   In order to fix the errors due to winding potentials and capacitances, the currents to be compared should be defined at identified or 'marked' terminals of the same polarity, when those terminals are at ground potential. (For convenience, each winding is fitted with a set of flexible 'standard leads' which are considered to be an integral part of the winding, the marked terminal being at the remote end of one of these leads.)

2   For repeatability of the measurements to be better than 0·1 parts per million, no direct current polarisation of the magnetic core is permissible.

3   The ambient field in the vicinity of the comparators must be low enough not to affect the measurement. Particular care must be taken with heavy current circuits. Each winding should be accessed by twisted-pair or coaxial leads and connections made at a 'star' point. A check of this effect can be made by disconnecting the current comparator, connecting its leads together, and

Fig. 1.3 *Construction of an early unshielded current comparator with ratios up to 60/ 5 amperes*

observing the voltage developed in the detection winding when full current is present in the remaining equipment and connecting leads.

These precautions were found to be essential when using the earlier comparators but, as refinements in design were introduced, they were found to have lesser impact.

A set of multiratio current comparators was then constructed,[4] covering ratios from 1/1 to 12/1 with a 5 ampere secondary rating. The ratio windings in these comparators were formed by six cables, all wound in one layer of seven turns each (Fig. 1.3). In order to study capacitance effects, however, in some of the comparators the primary and secondary winding conductors were intermixed in each cable, while in others whole cables were assigned alternately to each winding. This provided an interwinding capacitance of about 7000 picofarads in the first group of comparators and about 600 picofarads in the remainder.

These comparators were self-calibrated at 1/1 ratio and at all other ratios by a step-up technique where the ratio of the comparator being calibrated was equal to the sum of the ratios of the other two comparators. The circuit required that the secondary windings be connected in series, and corrections had to be applied for two non-standard conditions:

1 Owing to capacitance leakage, the current entering the unmarked terminal of a winding is not the same as that leaving the marked terminal.
2 Owing to the impossibility of maintaining all secondary winding marked terminals at ground potential, the component of error due to internal capacitance of one was shifted.

Methods for estimating and also measuring these corrections were established.

Attempts to extend these techniques to higher ratios were frustrated by the inability to continue with the single-layer-cabled ratio winding configuration. With multiple-layered windings it was not possible to effectively balance the magnetising forces within the windings themselves and the resulting leakage fluxes were responsible for a large increase in error.

The solution to this problem was found to be a magnetic shield[5] in the form of a hollow toroid which surrounded the magnetic core and detection winding, thus protecting them from the leakage fluxes of the ratio windings located outside the shield and from ambient magnetic fields. An early magnetic shield, machined from soft iron and silver plated, is shown in Fig. 1.4. Later shields were assembled from magnetic-tape-wound toroids and discs. With magnetic shielding, the error shifts corresponding to quite extreme non-uniformities in the ratio windings are held to one part in 10 million or less.

The final step in the evolution of the alternating current comparator was the

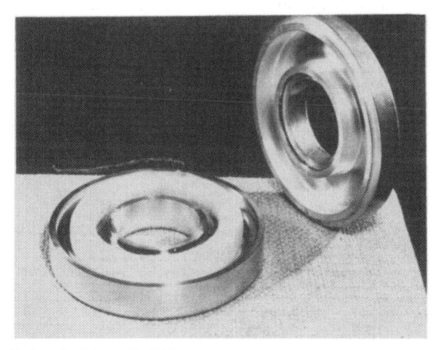

Fig. 1.4 *The first magnetic shield*

introduction of the compensation winding.[6] This winding is located inside the magnetic shield and has the same number of turns as one of the ratio windings outside the shield. When these two windings are connected in parallel, energy transfer between ratio windings is enabled by current transformer action using the magnetic shield as the core. Since the paralleled windings have the same number of turns, the current ratio function of the comparator remains relatively unimpaired.

1.3 The direct current comparator

Unlike three-winding voltage transformers, the current comparator operates with its magnetic core at zero flux. It can therefore be used for direct current ratio measurements using two-core magnetic modulation techniques to detect ampere-turn balance.[7] Except for the two-core detector, the configuration is essentially the same as for the alternating current comparator. The magnetic shield performs three functions:

1   It protects the detector cores from the leakage fluxes of the ratio windings outside the magnetic shield.
2   It impedes the current in the ratio winding circuits caused by voltages generated by the modulation circuitry.
3   It contributes to the dynamic response of the ratio balance by providing strong mutual coupling between the primary and secondary windings.

Since inductive coupling is not available for direct currents, it must be provided by electronic means. Accordingly, the output of the ampere-turn balance detector is used to control a slave current source supplying the secondary winding so as to bring about ampere-turn balance with the current in the primary ratio winding.

No equivalent of the compensation winding in alternating current comparators is available for the direct current comparator.

References

1 A. M. Thompson. 'The precise measurement of small capacitances.' *IRE Trans. Instrumentation*, **6–7**, December 1958, pp. 245–53
2 M. C. McGregor, J. F. Hersh, R. D. Cutkosky, F. K. Harris, and F. R. Kotter. 'New apparatus at the National Bureau of Standards for absolute capacitance measurement.' *IRE Trans. Instrumentation*, **6–7**, December 1958, pp. 253 61
3 I. Obradovic, P. Miljanic, and S. Spiridonovic. 'Prufung von Stromwandlern mittels eines Stromkomparators und eines electrischen Hilfssystems (Testing of current transformers with a current comparator and an auxiliary electrical system).' *ETZ(A)*, **78**(19), October 1957, pp. 699–701

4 N. L. Kusters and W. J. M. Moore. 'The current comparator and its application to the absolute calibration of current transformers.' *Power Apparatus and Systems*, 53, April 1961, pp. 94–104
5 P. N. Miljanic, N. L. Kusters, and W. J. M. Moore. 'The development of the current comparator, a high accuracy A-C ratio measuring device.' *Communications and Electronics*, 63, November 1962, pp. 359–68
6 N. L. Kusters and W. J. M. Moore. 'The compensated current comparator: a new reference standard for current transformer calibrations in industry.' *IEEE Trans. Instrumentation and Measurement*, **IM-13**(2–3), June–September 1964, pp. 107–14
7 N. L. Kusters, W. J. M. Moore, and P. N. Miljanic. 'A current comparator for precision measurement of D-C ratios.' *Communications and Electronics*, 70, January 1964, pp. 22–7

Chapter 2

Background and fundamental theory

2.1 Background

The process of electrical measurement consists of comparing the unknown against a standard. Often the standard is in the form of a spring torque, and many elegant schemes have been developed for the electromechanical conversion. The accuracy of such direct methods however is limited, and for more precise measurements the comparison is made against an artifact of like kind, using bridge circuits for scaling and null techniques. Either the same current is passed through the unknown and the standard, and the voltage drops across them are compared, or the same voltage is applied to both and the current ratio is measured. Many fine bridges have been developed in the past but all have problems which limit their range of applicability and the accuracy attainable. The task in measurement development is to devise methods of reducing the impact of these problems.

2.1.1 Direct current bridges
Until the advent of the current comparator, direct current bridges were constructed with resistors only. Typical examples are the Wheatstone bridge and the Kelvin double bridge. Some of the problems that may be encountered with such bridges are

1 Heating
2 Thermal electromotive forces
3 Lead resistances
4 Contact resistances.

Temperature rise in a resistance usually causes its value to change and care must be exercised to keep the current within allowable limits. This may govern the choice of bridge to be used. In bridge configurations where the voltage across adjacent arms is the same, the power loss will be determined by V^2/R and greater heating will occur in the lower-value resistor. If however, as in the

Kelvin bridge, the same current is passed through both resistors, the power loss will be given by I^2R and the higher-valued resistor will have the greater loss.

Thermal electromotive forces are generated by thermal gradients. Their magnitudes are reduced by controlling the temperature environment, by using bare copper strands in flexible conductors, and by using bare metal-to-metal contacts or, where necessary, low thermal solders. The residual effect of thermal electromotive forces is usually eliminated by measuring with both forward and reverse currents and averaging the two results.

Lead resistances must be considered in any accurate measurement and taken into account. In some bridges, this can be accomplished merely by measuring their resistances alone and subtracting these from the indicated value. The four-terminal resistor configuration provides the means for defining a resistance independent of the leads, but bridges used to measure such resistances (for example the Kelvin double bridge) require that a preliminary lead balance be made.

Contact resistances arise in switches and while, by design, these can be kept within specified limits, they are of indeterminate value. Their effect becomes very important in a multiple-decade array in which the contact resistances of the most significant digit decade may be more or less equal in value to the least significant digit of the array, thus limiting the overall resolution attainable.

2.1.2 Alternating current impedance bridges
Alternating current impedance bridges are constructed with resistors, capacitors, inductors, and mutual inductors. The most widely known are perhaps the Maxwell-Wien bridge and the Schering bridge, with several variations of the former being known by the name of the person who proposed them. Each has particular attributes that makes it preferable for certain measurements.

All of the problems associated with direct current bridges are present in alternating current impedance bridges except those due to thermal electromotive forces. In addition, there are three others of importance:

1   Component impurities leading to phase angle errors
2   Electrostatic interference
3   Electromagnetic interference.

The three-terminal gas-dielectric capacitor and an unloaded mutual inductor are the only components of an impedance bridge that can be considered to be free of phase angle error. Non-inductive resistors can be constructed but shunt capacitance is still present and difficult to control. Inductors and solid-dielectric capacitors also have loss components leading to phase defects. All of these must be taken into account in the bridge equation.

Electrostatic interference is caused by parasitic capacitive coupling between higher-potential and lower-potential portions of the circuit, and from the external circuitry. This problem is more or less eliminated by using appropriate shields. However, it may be necessary to drive these shields to a certain potential for them to be most effective, and this requires a second balance to be made.

Electromagnetic interference is caused by inductive coupling between currents in different branches of the bridge and from the external circuitry. The use of twisted go-and-return leads to each component from a central junction area so as to eliminate open loops between conductors is an effective solution to this problem. To ensure that the currents in both conductors in a twisted pair are of equal magnitude and that no current is lost through an unavoidable third terminal, the twisted pair can be passed several times through the window of a ferromagnetic toroid.

2.1.3 Alternating current transformer bridges

The use of a transformer to realise the two ratio arms of a bridge has several advantages. Since the ratio is based on turns only, it is extremely stable and independent of temperature. Contact resistances have a lesser effect and the resolution of the balance can be increased. Ratio variations caused by current loading depend on the leakage impedances only, which are very small; hence the shields required to prevent electrostatic interference can be connected directly to ground with negligible effect on the bridge balance.

It is relatively easy to construct a low-ratio transformer with a correspondence between the turns ratio and the current or voltage ratio that is better than one part in 1 million. It is not so easy however to realise the same accuracy with larger ratios involving a correspondingly larger ratio of conductor sizes and multiple-layer winding configurations. The leakage impedances become much larger and difficult to control. As a consequence, the ampere-turns of the winding are not uniformly imposed on the magnetic core and, in addition, the interwinding capacitive leakage currents that are driven by the impedance voltage drops have greater significance. In an effort to solve these problems, the concepts of induced voltage per turn, the ampere-turns acting on a magnetic core, and the meaning of the primary and secondary currents, had to be re-examined, and a new approach was sought which resulted in the invention of the current comparator. An additional feature of the invention was greatly increased protection against external electromagnetic interference.

2.1.4 Current transformers

Current transformers have long been used to provide the means of transforming currents from one magnitude level to another with very high ratio accuracy and phase correspondence. For many years the calibration of such devices has been obtained by reference to standard transformers using differential comparison techniques.

The standard transformer, which for convenience and cost is usually multi-ratio, must be very accurate if the need to apply corrections is to be avoided and measurement uncertainties are to be kept to a minimum. Many very good standard transformers have been developed for this application. Their errors are usually very small and stable, but nevertheless sensitive to the secondary load or burden. The errors of the current comparator however, because of its design features, are smaller and equally stable, and relatively insensitive to burden. This enables it to perform the duties of both a current ratio standard and a high-current supply transformer and, with the aid of an electronic amplifier, apply the desired burden to the current transformer being calibrated. It also provides the means to control circuit potentials automatically so that auxiliary grounding circuits are not required.

2.2 Ideas and concepts

A basic law of nature, named after one of the pioneers of electrical science, Ampère, states that the line integral of the magnetising force around a closed path is equal to the sum of the currents which are linked with that path:

$$\int_l (\boldsymbol{H} . \mathrm{d}\boldsymbol{l}) = \sum_s i \tag{2.1}$$

This equation is really the integral form of the basic Maxwell equation

$$\mathrm{curl}\, \boldsymbol{H} = \boldsymbol{J} + \frac{\mathrm{d}\boldsymbol{D}}{\mathrm{d}t} \tag{2.2}$$

The sum $\sum i$ in Ampère's law (2.1) includes conductive as well as dielectric currents ($\mathrm{d}\boldsymbol{D}/\mathrm{d}t$ in (2.2)). At power and audio frequencies, and with the actual electric fields encountered in current comparators, the dielectric currents are usually much smaller than the conductive currents and their effect may be neglected.

The general idea of the current comparator measuring technique may be grasped from Fig. 2.1. If all capacitive currents are neglected it can be stated that

$$\sum i = N_1 i_1 - N_2 i_2 \tag{2.3}$$

and according to (2.1)

$$N_1 i_1 - N_2 i_2 = \int_l (\boldsymbol{H} . \mathrm{d}\boldsymbol{l}) \tag{2.4}$$

Suppose that means are available to measure the line integral $\int_l (\boldsymbol{H} . \mathrm{d}\boldsymbol{l})$ and also to adjust one of the currents, say i_2, until the line integral reaches zero value. Equation (2.3) will then state that the current ratio is exactly

$$\frac{i_1}{i_2} = \frac{N_2}{N_1} \tag{2.5}$$

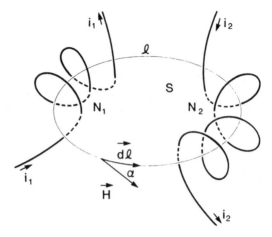

Fig. 2.1 *Implementation of Ampère's law*

Measurements of the line integral $\int_l (\boldsymbol{H} . \mathrm{d}\boldsymbol{l})$ itself is, however, a challenging if not impossible task. If the currents to be compared are direct currents, some sort of magnetic field sensing device has to be used, as for example the second-harmonic flux-gate magnetometer. The task of obtaining an indication that line integral $\int_l (\boldsymbol{H} . \mathrm{d}\boldsymbol{l})$ is equal to zero is much simpler if alternating currents are being compared.

Figure 2.2 shows a possible approach for an alternating current comparison.[1] A uniformly distributed detection coil is wound around the line *l*. It is supposed that all of the turns are of small diameter and the same cross-section.

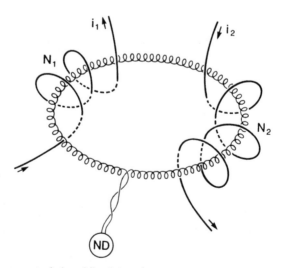

Fig. 2.2 *Measurement of closed line integral*

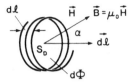

Fig. 2.3 *Element of the detection winding*

Owing to the alternating currents the magnetic field along the line *l* changes and induces a voltage in the detection winding. Depending upon the space geometry of the wires carrying the currents i_1 and i_2, this induced voltage in some sections of the detection winding may have a positive sign and in some sections a negative sign. But if the relation $N_1 i_1 - N_2 i_2 = 0$ is fulfilled, the total induced voltage appearing at the terminals of the voltage null detector ND should be zero.

This conclusion is based upon the following assumptions:

1   The cross-sections S_D of all the turns of the detection winding are the same.
2   The local magnetic field is uniform, that is it has the same value in the close neighbourhood of the point on line *l* which is the axis of the detection winding (see Fig. 2.3).
3   The detection winding has a large number of turns uniformly distributed with a turns density $n = N/l$.

Under these assumptions the voltage induced in the element d*l* of the detection winding is

$$dv = n \, dl \frac{d\Phi}{dt} = n \, dl \frac{d}{dt} (B S_D \cos \alpha) \tag{2.6}$$

or

$$dv = n S_D \mu_0 \frac{d}{dt} (H \, dl \cos \alpha) \tag{2.7}$$

The total induced voltage in the detection winding will be

$$v = \int_l dv = \int_l n S_D \mu_0 \frac{d}{dt} (H \, dl \cos \alpha) \tag{2.8}$$

or

$$v = n S_D \mu_0 \frac{d}{dt} \int_l (\boldsymbol{H} \cdot d\boldsymbol{l}) = n S_D \mu_0 \frac{d}{dt} \sum i \tag{2.9}$$

since the permeability of free space μ_0, the detection winding cross-section S_D, and the turns density *n* may be brought out from the integral as constants.

2.2.1 Sensitivity

The sensitivity of the suggested current comparator shown in Fig. 2.2, putting aside temporarily the accuracy problems associated with the above assumptions, can now be investigated. Suppose that the detection winding cross-section $S_D = 1 \text{ cm}^2$ and the turns density $n = 4$ turns/mm. For sinusoidal currents the sensitivity, expressed in ohms as the ratio of induced voltage per unbalanced current ampere-turns, will then be

$$Z = \frac{V_D}{\Sigma I} = j\omega n S_D \mu_0 \tag{2.10}$$

or

$$|Z| = 2\pi n S_D \mu_0 f \tag{2.11}$$

Note that the sensitivity (2.11) does not depend on the toroid diameter or on the positions of windings carrying the currents to be compared.

For this configuration at a frequency of 60 Hz the sensitivity is

$$|Z| = 2\pi(4 \times 10^3)(10^{-4})(4\pi \times 10^{-7})(60) \approx 0\cdot2 \text{ mV/A}$$

It is obvious that at 60 Hz this sensitivity is very small. For example if the ampere-turns of the compared current are 100 AT, then a 1 part per million (ppm) difference would be 100 μAT, and this difference will bring about a detection winding terminal voltage of only 20 nV.

The obvious step to increase the current comparator sensitivity is to introduce a high-permeability material inside the detection winding.[2] A relative permeability μ_r greater than 10 000 is readily available in presently available alloys. The sensitivity will increase in proportion according to

$$|Z| = 2\pi n S_D \mu_r \mu_0 f \tag{2.12}$$

For the same configuration and a magnetic core having a relative permeability $\mu_r = 50\,000$, a 1 ppm difference will be seen as a 1 mV induced voltage, a comfortable value with respect to the noise level and available voltage detectors.

2.2.2 The magnetic error of a current comparator

Introducing high-permeability material solves the sensitivity problem, but another more serious one is posed. Because the relative permeability μ_r is *not* constant for various parts of the magnetic core, μ_r *cannot* be brought out from the integral in eqn. (2.8). Consequently a zero indication of the null detector does not mean that the sum of all currents passing through the detection winding opening is zero:

$$v = \frac{d}{dt} \int_l n S_D \mu_r \mu_0 H \, dl \cos \alpha \neq n S_D \mu_r \mu_0 \frac{d}{dt} \sum i \tag{2.13}$$

This fact is responsible for the magnetic error of a current comparator.

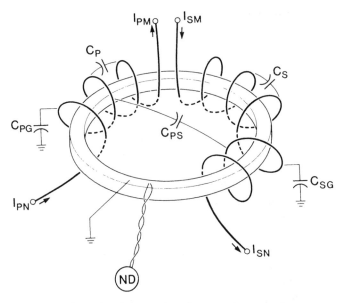

Fig. 2.4 *Capacitances influencing the operation of a current comparator*

2.2.3 *The capacitive error of a current comparator*

A current comparator without the magnetic error would be a true ampere-turn balance detector, but this does not imply that the mere placing of turns on such a comparator would make it a true current ratio device. Capacitive leakage may prevent the ampere-turns imposed on the core from equalling the product of the total turns of a winding and the current measured at either of its terminals.

Figure 2.4 illustrates the problems concerning the stray capacitive currents. Some small portion of conductor current may leave its path and reach the ground through the winding-to-ground capacitances C_{PG} and C_{SG}; jump from one turn to another of the same winding through turn-to-turn capacitances C_P and C_S; or jump from primary to secondary turns through the interwinding capacitance C_{PS}. Thus in high-precision measurements the notion of terms such as the primary or secondary current is only an approximation. The primary and secondary currents have to be specified at their terminals (here M and N) because the current which enters a winding is not the same as the current which leaves it. The actual ampere-turns of the primary and secondary windings are obviously not the product of the number of turns and the currents at either of the terminals. The definition of current ratio is even more complex because the capacitive currents depend upon the winding potentials.

Capacitive currents are responsible for the capacitive error of a current comparator.

2.3 Definition of the error of a current comparator

The error of a current comparator may be expressed most conveniently by

$$\varepsilon = \varepsilon_p + j\varepsilon_q = \frac{N_S I_S}{N_P I_P} - 1 \tag{2.14}$$

where

ε_p, ε_q = in-phase and quadrature components of the complex error ε
N_P, N_S = number of primary and secondary turns, respectively
I_P, I_S = primary and secondary complex currents, respectively.

The conditions implied by this definition of comparator error are not sufficient, however, for measurements where the extremely high accuracies of 1 ppm or better are desired. The complete equivalent circuit of a current comparator should take into account not only the magnetic phenomena, but also the self-capacitances of the winding and capacitive couplings between the various windings and to ground.

Thus, when the error of a current comparator is specified, it is necessary to state at which terminals the currents are being measured, and also what potential conditions exist in the windings during the measurement.

The origin of current comparator error is entirely distinct from that of current transformer error. Measurements with a comparator are taken when the fundamental frequency voltage across the detection winding is zero; that is, when the average flux along the magnetic core path is zero. The error of a current transformer, which results from the magnetising and core-loss currents associated with the magnetic flux required for energy transfer, is not present in the current comparator. This fact alone explains the great improvement in accuracy. However, certain residual causes of error, existent but often overlooked in current transformers, still remain and must be overcome to achieve accuracies of 1 ppm or better.

The error of a current comparator may be divided into two components:

$$\varepsilon = \varepsilon_m + \varepsilon_c \tag{2.15}$$

where ε_m is the magnetic error and ε_c is the capacitive error. As noted earlier, magnetic errors arise from the effects of the leakage fluxes owing to the lack of symmetry in the winding combined with the non-uniformities in the detection winding and, more especially, in the magnetic core permeability. The capacitive errors may be traced to the various capacitive couplings.

2.4 Errors of magnetic origin

In a current comparator, the measurement is made when the sum of the effects of the various currents on the detection winding is zero. If the primary and secondary windings could be made to coincide exactly in space at ampere-turn balance, the magnetic field produced by the two currents would be zero everywhere. Such a comparator would be free from magnetic error, no matter how non-uniform the core or detection winding might be.

In practice, such a winding design may be approached but never fully realised. In the early stage of current comparator development, to minimise the leakage fluxes, the primary and secondary windings were wound together in the form of one cable with several strands.[3] Some of the strands were then connected in series and some in parallel to make the secondary and primary windings. Extension of that design to higher ratios was not practical for physical reasons and, furthermore, the inherently large capacitance between the windings presented additional problems relating to comparator accuracy.

In current comparators where the primary and secondary windings are physically separated, a leakage flux is inevitable. Moreover, it is almost impossible to confine the leakage flux in the space between the primary and secondary windings, even with the most careful design of corresponding toroidal windings. Some part of the leakage flux, owing to the attraction from the high permeability of measuring core and the inevitable winding asymmetry, will always reach the core and cause a magnetic error.

Meaningful experimental results show how the leakage flux induces magnetic error in the current comparator.[4] A toroidal magnetic core is surrounded by a magnetic field created by alternating current in a long, thin, rectangular probe coil, perpendicularly located in the centre of the toroid (Fig. 2.5). Wound on the

Fig. 2.5 *Investigation of the effect of dipole leakage flux*

Table 2.1 *Voltages induced in section α of toroid as a function of probe-coil position β*

α, degrees	β, degrees											
	0	15	30	45	60	75	90	105	120	135	150	165
0	1·29	1·23	1·08	0·87	0·58	0·25	−0·08	−0·42	−0·77	−0·99	−1·17	−1·27
30	1·10	1·20	1·25	1·20	1·05	0·84	0·56	0·24	−0·12	−0·43	−0·74	−0·93
60	0·65	0·90	1·08	1·20	1·21	1·16	1·02	0·82	0·52	0·21	−0·09	−0·38
90	0·03	0·31	0·60	0·86	1·03	1·16	1·19	1·15	1·01	0·83	0·57	0·26
120	−0·58	−0·28	−0·03	0·29	0·57	0·82	1·00	1·15	1·18	1·14	1·03	0·82
150	−1·02	−0·83	−0·60	−0·32	−0·04	0·27	0·52	0·80	1·01	1·14	1·19	1·17
180	−1·23	−1·17	−1·06	−0·88	−0·62	−0·35	−0·07	0·23	0·55	0·81	1·00	1·17
210	−1·06	−1·20	−1·27	−1·22	−1·08	−0·91	−0·67	−0·38	−0·06	0·24	0·54	0·85
240	−0·60	−0·92	−1·10	−1·24	−1·30	−1·26	−1·11	−0·93	−0·63	−0·34	−0·05	0·27
270	0·03	−0·33	−0·67	−0·95	−1·14	−1·28	−1·33	−1·28	−1·10	−0·90	−0·63	−0·36
300	0·66	0·33	−0·03	−0·36	−0·70	−0·98	−1·16	−1·30	−1·34	−1·28	−1·10	−0·95
330	1·12	0·91	0·60	0·27	−0·07	−0·41	−0·73	−0·98	−1·20	−1·30	−1·33	−1·27
All sections in series	0.37	0.14	−0.10	−0.30	−0.52	−0.71	−0.84	−0.94	−0.96	−0.90	−0.80	−0.59

Measured in mv rms (root mean square). Negative sign indicates phase reversal. Twelve 20-turn sections on 12·5-cm. ID by 14·0-cm. OD by 1·25-cm. height HyMu 80 toroidal core; 3·8-cm wide probe coil; 10 ampere turns at 60 Hz.

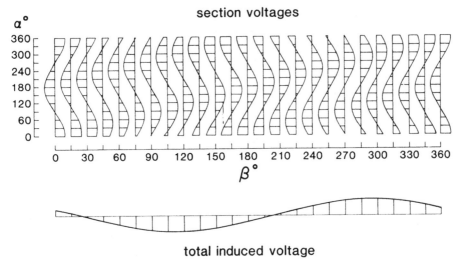

Fig. 2.6 *Voltage induced by dipole leakage flux*

core is a detection winding which is divided into 12 sections, each of which can be connected in series with the others or measured independently. The sum of the voltages induced in each section will not necessarily be zero, in spite of the obvious fact that sum of currents passing through the toroid opening is zero ($\Sigma I = 0$). Table 2.1 shows a typical set of voltages induced in the individual sections, designated by co-ordinate α, as a function of the orientation of the probe coil, co-ordinate β. The same results are shown graphically in Fig. 2.6.

The following conclusion may be drawn from the results given in Table 2.1:

1   Symmetry of magnetic flux inside the toroidal core, with respect to the magnetic dipole or probe coil, does not exist. This is indicated by the fact that the voltages in individual diametrically opposed sections symmetrically located with respect to the probe coil are not equal, and consequently the sum of all these voltages for each position β of the probe coil is not zero.
2   Symmetry of the magnetic field outside the toroidal core, with respect to the magnetic dipole, does exist. This is evident from the fact that the voltage difference between two adjacent sections is the same as that between two adjacent sections that are diametrically opposite and symmetrical with respect to the magnetic dipole. This voltage difference is proportional to the flux entering the toroid at the point.
3   The voltage induced in a uniformly distributed winding on the toroid approximates a sinusoidal function of the dipole orientation β.

To explain these results, it must be noted that, first, the core is constructed from magnetic material of very high initial permeability ($\mu/\mu_0 \geqslant 50\,000$) and, second, that core cross-section is small compared with the toroid diameter (inside diameter/outside diameter (ID/OD) $\geqslant 0.9$).

From the first it follows that, regarding the magnetic field distribution outside the core, it may be concluded that the magnetic potential is almost the same everywhere in the close vicinity of the core. If this were not so, the flux density inside the core would be enormous, since the effective core permeability is very high. The distribution of the magnetic field ouside the core then does not depend on the non-uniformities of the core's magnetic characteristics, but only on its geometry and that of the probe coil. So, if the magnetic dipole is symmetrically located in the middle of the core, the external magnetic field is bound to be symmetrical with respect to the magnetic dipole, which is in accordance with experimental results.

The magnetic flux distribution inside the core is an entirely different problem. Perhaps the best approach to understand magnetic field distribution inside the core is to use the magnetic circuit concept. This is possible since the ratio of the inner to outer diameters of the core is close to unity and the space problem can therefore be reduced to one dimension.

Leakage flux which enters at some point into the core splits into two parts. How large these parts are relative to one another depends upon the uniformity of core reluctance. If the reluctance is the same for the left and right side of the core, the leakage flux will divide into two equal parts. But if, for example, the left side of the core has a higher reluctance than the right side, the larger part of the leakage flux will pass through the right side.

A rough picture of the magnetic field is shown in Fig. 2.7. If the total field

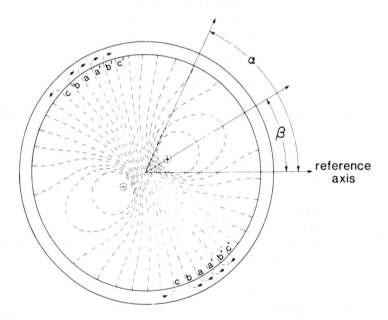

Fig. 2.7 *Distribution of magnetic field outside and inside magnetic toroid*

were symmetrical, lines *a*, *b* and *c* would close through the left half of the toroid while lines *a'*, *b'* and *c'* would close through the right half. Actually, the location of the divergence is shifted, and the line *c* closes to the left while *b*, *a*, *a'*, *b'* and *c'* close to the right.

From the magnetic circuit point of view eqn. (2.8) may be rewritten, and for the sinusoidal condition the induced voltage of the detection winding will be given by

$$V_D = j\omega \int_l n\Phi \; dl \qquad (2.16)$$

with

$$\int_l \Phi \; dR_m = \int_l \frac{\Phi}{\mu S_D} \; dl = \sum I \qquad (2.17)$$

where

ω = angular frequency
n = turns density
Φ = flux in the magnetic circuit
dR_m = element of magnetic circuit reluctance
μ = effective permeability
S_D = magnetic circuit cross-section
dl = element of the mean magnetic path
$\sum I$ = total magnetomotive force acting upon the magnetic circuit, i.e. sum of all currents passing through the toroidal opening.

Some of the quantities in eqns. (2.16) and (2.17) are complex. The magnetic flux is not in time phase with the magnetomotive force owing to the core losses, and the magnitude and phase relation between them depends on the characteristics of the magnetic material. For simplicity, this phenomenon is considered to be completely resolved by attributing a complex property to the effective permeability or to the reluctance. The latter quantities therefore should be understood to be merely complex coefficients which express the fundamental frequency relation between the magnetomotive force and the magnetic flux.

If the detection winding density *n*, winding cross-section S_D, and complex permeability μ were constant, eqns. (2.16) and (2.17) would give

$$V_D = j\omega\mu n S_D \sum I \qquad (2.18)$$

and the induced voltage in the detection winding would be proportional to the total ampere-turns. Experiments however do not agree with eqn. (2.18), as was pointed out earlier.

Investigating the various factors and, after performing many experiments which included rewinding the same core many times, an opinion was reached that winding density and effective cross-section area are only slightly responsible for the non-uniformity. The conclusion derived from these tests was that the only possible explanation for the existence of a detection winding voltage, when the sum of currents through the toroid window is zero, is that the effective permeability is not constant along the magnetic core path. The assumption that μ varies by 10% is, in fact, sufficient to explain this result.

To conclude, owing to the combined action of the leakage fluxes and core non-uniformity, eqn. (2.18) is not valid, and the following expression for the induced voltage of the detection winding must be used instead:

$$V_D = j\omega(n\mu S_D)_{av}\left(\sum I + I_0\right) \qquad (2.19)$$

In the current comparator

$$\sum I = N_P I_P - N_S I_S$$

$$I_0 = \varepsilon_m N_P I_P$$

where ε_m is by definition the magnetic error of the comparator.

2.4.1 Correction winding

Compensation for the non-uniformity of the core along its magnetic path may be realised, in part, by suitable variations in the winding density aiming to keep the product $n\mu S_D$ constant. The permeability is a complex quantity, however, while winding density is scalar, and hence complete compensation cannot be achieved.

Adjustment of the winding density to produce optimum compensation for the variation in effective permeability is tedious and unwarranted by the results obtained. A fairly good correction however can be achieved by the following procedure:

1 Using the probe coil shown in Fig. 2.5, determine the position β which produces the maximum voltage in the detection winding V_{Dmax}.
2 Using a temporary coil located on the toroid at the position where the plane of the probe coil intercepts the toroid ($\alpha = \beta$), determine the induced voltage per turn V_{D0}.
3 Over that part of the toroid of which the location of the temporary coil was the centre, wind N turns, uniformly distributed, where

$$N = \frac{\pi}{2}\frac{V_{Dmax}}{V_{D0}} \qquad (2.20)$$

4 Connect the additional turns in series with the main detection winding, in such a sense that the overall variation in the detection winding voltage caused by change in the probe position is a minimum.

Regarding the contribution of the correction winding for improving the accuracy, it should be emphasised that, owing to the unsymmetry of ratio windings, the actual leakage flux distribution is not known. Thus the correction for the probe leakage flux does not imply that it will work as successfully in a real measurement with, most probably, a completely different leakage flux distribution.

It must be emphasised also that a correction winding designed for one frequency may not be applicable for any other frequency.

2.4.2 Copper shield

To avoid all kinds of capacitive coupling and to reduce unwanted voltages derived from the environment, a copper shield over the detection winding is essential. A thin copper shield, connected to the ground side of the detection winding, is completely effective in reducing electrostatic interference. Needless to say, the shield should not be constructed in such a way that it constitutes a short-circuited turn around the core. A slit at one of the edges or an insulated overlap provides the necessary break.

As is well known, a copper shield repulses the magnetic flux lines by virtue of induced eddy currents.[5] It was thought that the copper shield could also prevent the leakage flux from reaching the measuring core and cause magnetic error. So to keep the leakage flux away from the detection winding a shield was made in the form of a heavy copper box. Unfortunately the actual shielding effect at low frequencies (50–60 Hz) was found to be so small that the copper shield alone gave no significant improvement in the reduction of magnetic error.

2.4.3 Magnetic shield

A breakthrough in the development of current comparators was the introduction of magnetic shielding.

In a current comparator, the terms 'working flux' and 'leakage flux' may be applied, respectively, to that part of the total magnetic flux whose lines *are*, or *are not*, entirely enclosed by the detection winding. The total voltage in this winding is then the sum of the voltages induced in its turns by the working and leakage flux. When the total current through the toroid window is zero, the working flux is zero. The leakage flux remains, however, and this may have its origin in the currents linking the toroid (primary and secondary windings) as well as in the currents not linking the toroid (external magnetic fields). The working flux is proportional to the term ΣI of eqn. (2.19), while the leakage flux corresponds to the term I_0.

The magnetic error of the current comparator can be reduced appreciably if leakage fluxes are kept from reaching the detection winding by surrounding the magnetic core and detection winding with a heavy magnetic shield.[4] The primary and secondary ratio windings are located outside the shield (Fig. 2.8).

The magnetic shield does not change the comparator sensitivity, but it will act as a shunt which decreases the passage of leakage fluxes through to the

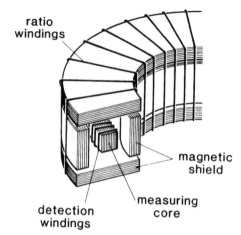

Fig. 2.8   *Magnetic shield surrounding the measuring core*

measuring core, thus greatly reducing the detection winding voltage when $\Sigma I = 0$. The shield should therefore have a high reluctance in directions perpendicular to the core, and a low reluctance in direction parallel with it.

Except for only a few geometrically simple enclosures, such as spherical and cylindrical shells, the analytical calculation of shielding effects is extremely complex. Using the field theory approach, based on Maxwell equations, and by satisfying the boundary conditions only general estimates can be obtained. For the practical problem, as it is in the investigation of shielding effects in current comparators, useful analytical results cannot be obtained. Nevertheless, for low frequencies, the magnetic circuit approach may give some coarse estimates of shielding effects and some practical recommendations for shield design.

Consider the measuring core and the shield which surrounds it as two concentric toroids. If the leakage flux enters this magnetic structure at one point and leaves it at another 180 degrees apart, to a first approximation it may be said that this flux splits into two halves. Now, if inside/outside toroid diameter ratio is not too small, the toroid may be substituted by a pair of two concentric cylinders. The flux density distribution problem may then be further reduced from three dimensions to two as shown in Fig. 2.9a.

To find the flux distribution, assume that

1 Almost all of the leakage flux passes through the shield and only a minor part through the measuring core.
2 The magnetic potential is zero everywhere in the measuring core since the core reluctance is small ($\mu/\mu_0 \gg 1$) and the magnetic flux is negligible.

From Fig. 2.9a it follows that the magnetomotive force acting upon the air gap is a function of the distance x as follows:

$$M = \Phi_L R_m \frac{x}{l} \qquad (2.21)$$

where Φ_L is one-half of the leakage flux and R_m is the shield reluctance of the half-toroid:

$$R_m = \frac{l}{\mu_s ac} \qquad (2.22)$$

The element of flux reaching the measuring core is

$$d\Phi'_L = \frac{M}{b} \mu_0 c \, dx = \Phi_L \frac{\mu_0}{\mu_s} \frac{x \, dx}{ab} \qquad (2.23)$$

and the total leakage flux through the measuring core is

$$\Phi'_L = \int_0^{l/2} d\Phi'_L = \frac{1}{8} \Phi_L \frac{\mu_0}{\mu_s} \frac{l^2}{ab} \qquad (2.24)$$

The attenuation factor or shield effectiveness A is defined as the ratio of the magnetic leakage flux which would impinge on the measuring core without the magnetic shield to that which would actually reach the measuring core with the shield installed:

$$A = \frac{\Phi_L}{\Phi'_L} = \frac{\mu_s}{\mu_0} \frac{8}{l^2} ab \qquad (2.25)$$

where

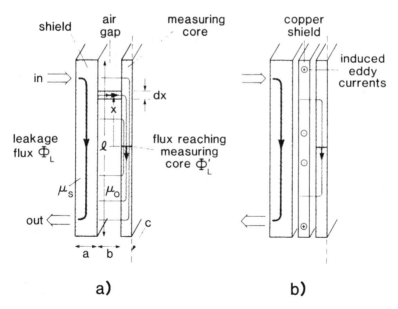

a)   **b)**

Fig. 2.9 *Attenuation of leakage flux penetration into measuring core by magnetic and copper shields*

μ_s = permeability of the magnetic shield
μ_0 = permeability of the air gap
   a = thickness of the shield
   b = air gap
   l = distance between the entrance and the exit of the leadage flux

From eqn. (2.25) it may be concluded that the shield effectiveness depends upon the ratio μ_s/μ_0; that it is higher for local leakage fluxes (l small); and that it depends on the product of the shield thickness and the air gap. So if the available space is given ($a + b$ fixed), a and b should be equal to get optimum shielding. This optimum is not however sharp, and it is quite tolerable to have the air gap even twice as large as the shield thickness or vice versa. Usually, when the detection winding, electrostatic shield, and deviation winding are placed on the measuring core, and the thick insulation needed to reduce capacitances is used between them, a satisfactory air gap is realised.

As has been mentioned earlier, even a heavy copper box only slightly changes the distribution of the leakage flux. But if a copper shield is inserted between the magnetic shield and measurement core it becomes effective, especially for radial leakage fields. The physical process is that the leakage flux induces eddy currents in the copper which produce opposing magnetomotive forces, and in this way the leakage flux is prevented from reaching the measuring core. One example of how eddy current paths are formed and how these currents reduce the leakage through the measuring core is shown in Fig. 2.9b.

The magnetic shield, as well as the electrostatic copper shield, should be constructed in such a way that the conductance around the measuring core is so small that the current which flows in the shield cannot, under any conditions, produce ampere-turns through the toroid window. Suitable gaps or insulation should be included to prevent current flow around the core.

2.5 Errors of capacitive origin

A measuring core with a uniformly wound detection winding, appropriately shielded, which is connected to a sensitive voltage detector, is actually a true ampere-turn null detector. This device should be able to detect a current of, for example, 1 μA passing through the toroid opening and at the same time be immune to a current of several hundreds of amperes in a cable passing and returning through the same opening.

To compare heavy currents such a device is almost perfect. For example, to measure current ratios of 100 A to 10 A it suffices to use one turn as the primary and ten turns as the secondary winding. The leakage capacitances will be completely negligible and the ampere-turns will be the same as current times the number of turns. When currents to be compared are small, however, and when even the primary has to have a large number of turns to obtain

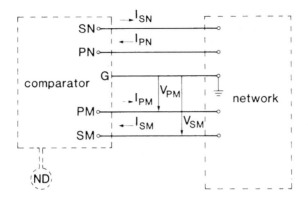

Fig. 2.10 *Connection of the current comparator to the external network*

satisfactory sensitivity, the capacitance currents may be significant. The total current passing through the toroid opening is no longer equal to the product of the amperes and the turns.[6,7,8]

In general, if the current comparator and its null indicator are suitably shielded from the ambient magnetic and electric fields, then five connections between the comparator and the rest of the network can be made at the points designated as PN, PM, SN, SM and G in Fig. 2.10.

The first approach to define the error of a current comparator is as the difference between the current ratio and the turns ratio (see (2.14)):

$$\frac{I_S}{I_P} = \frac{N_P}{N_S}(1 + \varepsilon) \tag{2.26}$$

This lacks precision because it uses the concept of primary and secondary currents. It is well known that owing to the capacitive currents between the windings, and between the windings and ground, the current entering terminal N is not necessarily the same as the current leaving terminal M. So for high-accuracy measurements the subsequent four current ratio errors may be defined as follows:

$$\frac{I_{SN}}{I_{PN}} = \frac{N_P}{N_S}(1 + \varepsilon_{NN}) \tag{2.27}$$

$$\frac{I_{SM}}{I_{PN}} = \frac{N_P}{N_S}(1 + \varepsilon_{NM}) \tag{2.28}$$

$$\frac{I_{SN}}{I_{PM}} = \frac{N_P}{N_S}(1 + \varepsilon_{MN}) \tag{2.29}$$

$$\frac{I_{SM}}{I_{PM}} = \frac{N_P}{N_S}(1 + \varepsilon_{MM}) \tag{2.30}$$

where

I_{PN}, I_{PM}, I_{SN}, I_{SM} = currents at the terminals PN, PM, SN and SM, respectively

N_P, N_S = number of turns in the primary and secondary windings, respectively

ε_{NN}, ε_{NM}, ε_{MN}, ε_{MM} = current comparator errors defined by these equations

Each of these errors ε_{NN}, ε_{NM}, ε_{MN} and ε_{MM} is dependent not only on the characteristics of the comparator but also on the voltages to ground at terminals PM and SM. Therefore in high-precision current ratio measurements the voltages to ground of two of the terminals must be stated. Usually current comparator error is given by eqn. (2.30), but with the necessary addition that I_P and I_S are defined as the currents at the marked terminals M of the primary and secondary windings when these terminals are at ground potential.

2.5.1 Capacitive error calculation

If the capacitive current arrives at the primary marked terminal without passing through all of the primary turns, the error is negative because it increases the amount of primary current without a corresponding effect on the current null detector. If the capacitive current passes through some primary turns but does not reach the primary marked terminal, the error is positive because the secondary current must be increased to balance the null detector. For the same reasons, if a capacitive current which passes through the secondary marked terminal does not pass through all secondary turns the error is positive, but if a capacitive current passes through some secondary turns but does not pass through the secondary marked terminal the error is negative.

For a reasonably uniform voltage and capacitance distribution, equations can be developed which allow a fairly good estimate of a current comparator's capacitive errors.[9] As an example, assume that both secondary and primary windings are uniformly wound, each in a single layer around the toroid, with a distributed capacitance per turn considered to be constant. Assume as well that

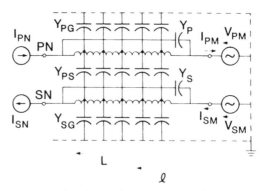

Fig. 2.11 *Capacitances in uniformly distributed primary and secondary windings*

Table 2.2   *Symbols for capacitive error calculation*

R_P	AC resistance of the primary winding
R_S	AC resistance of the secondary winding
L_P	leakage inductance of the primary winding
L_S	leakage inductance of the secondary winding

$$Z_P = R_P + j\omega L_P$$
$$Z_S = R_S + j\omega L_S$$

C_P	capacitance in shunt with the primary winding
C_{PG}	capacitance between the primary winding and ground
C_{PS}	capacitance between the primary and secondary windings
C_S	capacitance in shunt with the secondary winding
C_{SG}	capacitance between the secondary winding and ground

$$Y_P \simeq j\omega C_P \qquad Y_{PG} \simeq j\omega C_{PG} \qquad Y_{PS} \simeq j\omega C_{PS}$$
$$Y_S \simeq j\omega C_S \qquad Y_{SG} \simeq j\omega C_{SG}$$

$n = N_S/N_P$	turns ratio
l	distance from the marked terminal
L	total length

all four terminals are brought out at the same place and, for the sake of generality, that the marked terminals are not at ground potential. Then the equivalent current comparator ratio winding circuit may be illustrated as in Fig. 2.11 (for symbols see also Table 2.2).

The actual secondary-to-primary marked terminal current ratio when the comparator is in balance, that is, when the null detector indicates zero, may be obtained by summing all of the ampere-turns. If the magnetic error of the current comparator is negligible, this sum should be zero ($\Sigma I = 0$). The ampere-turns are as follows:

1   The primary ampere-turns which would be imposed on the core if the capacitive currents were absent:

$$+ N_P I_P \tag{2.31}$$

2   The ampere-turn loss due to the primary winding voltage drop and the shunt capacitance current bypassing that winding:

$$- N_P Z_P I_P Y_P \tag{2.32}$$

3   The positive ampere-turns due to the primary winding voltage drop and

the leakage capacitive current from the primary winding to ground passing through $(1 - l/L)N_P$ turns:

$$+ \int_0^L \frac{N_P}{L}(L - l)Z_P I_P \frac{l}{L} \frac{Y_{PG}}{L} \, \mathrm{d}l \tag{2.33}$$

4   The positive ampere-turns due to the primary and secondary winding voltage drops and the capacitive current from the primary to the secondary winding passing through $(1 - l/L)N_P$ turns:

$$+ \int_0^L \frac{N_P}{L}(L - l)(Z_P I_P + Z_S I_S) \frac{l}{L} \frac{Y_{PS}}{L} \, \mathrm{d}l \tag{2.34}$$

5   The secondary ampere-turns which would be imposed on the core if the capacitive currents were absent:

$$- N_S I_S \tag{2.35}$$

6   The ampere-turn gain due to the secondary winding voltage drop and the shunt capacitance current bypassing that winding:

$$+ N_S Z_S I_S Y_S \tag{2.36}$$

7   The negative ampere-turns due to the secondary winding voltage drop and the leakage capacitance current from the ground to the secondary winding passing through $(1 - l/L)N_S$ turns:

$$- \int_0^L \frac{N_S}{L}(L - l)Z_S I_S \frac{l}{L} \frac{Y_{SG}}{L} \, \mathrm{d}l \tag{2.37}$$

8   The negative ampere-turns due to the primary and secondary winding voltage drops and the capacitive current from the primary to the secondary winding passing through $(1 - l/L)N_S$ turns:

$$- \int_0^L \frac{N_S}{L}(L - l)(Z_P I_P + Z_S I_S) \frac{l}{L} \frac{Y_{PS}}{L} \, \mathrm{d}l \tag{2.38}$$

9   The negative ampere-turns due to the primary winding voltage and the capacitive current passing to ground through $(1 - l/L)N_P$ turns:

$$+ \int_0^L \frac{N_P}{L}(L - l)V_{PM} \frac{Y_{PG}}{L} \, \mathrm{d}l \tag{2.39}$$

10   The ampere-turns due to the primary winding voltage and the capacitive current from the primary winding to the secondary winding, being

positive for $N_P(1 - l/L)$ turns of the primary winding and negative for $N_S(1 - l/L)$ turns of the secondary:

$$+ \int_0^L \left(\frac{N_P}{L} - \frac{N_S}{L} \right)(L - l)V_{PM} \frac{Y_{PS}}{L} \, dl \tag{2.40}$$

11   The positive ampere-turns due to the secondary winding voltage and the capacitive current passing to ground through $N_S(1 - l/L)$ turns:

$$+ \int_0^L \frac{N_S}{L}(L - l)V_{SM} \frac{Y_{SG}}{L} \, dl \tag{2.41}$$

12   The ampere-turns due to the secondary winding voltage and the capacitive current from the secondary winding to the primary winding, being positive for $N_S(1 - l/L)$ turns of the secondary and negative for $N_P(1 - l/L)$ turns of the primary:

$$+ \int_0^L \left(\frac{N_S}{L} - \frac{N_P}{L} \right)(L - l)V_{SM} \frac{Y_{PS}}{L} \, dl \tag{2.42}$$

Summing all these terms and equalising to zero ($\Sigma I = 0$), after integration and rearrangement the following equation is obtained:

$$\begin{aligned}
N_S I_S &\left\{ 1 + Z_S \left[-Y_S + \frac{Y_{SG}}{6} + \left(1 - \frac{1}{n} \right) \frac{Y_{PS}}{6} \right] \right. \\
&\left. + \frac{V_{SM}}{I_S} \left[\frac{Y_{SG}}{2} + \left(1 - \frac{1}{n} \right) \frac{Y_{PS}}{2} \right] \right\} \\
&= N_P I_P \left\{ 1 + Z_P \left[-Y_P + \frac{Y_{PG}}{6} + (1 - n) \frac{Y_{PS}}{6} \right] \right. \\
&\left. - \frac{V_{PM}}{I_P} \left[\frac{Y_{PG}}{2} + (1 - n) \frac{Y_{PS}}{2} \right] \right\}
\end{aligned} \tag{2.43}$$

If the terms of higher order are neglected ($1/(1 \pm x) \simeq 1 \mp x$ if x is small), the current comparator error due to capacitive currents is

$$\begin{aligned}
\varepsilon_{C.MM} &= Z_S \left[Y_S - \frac{Y_{SG}}{6} - \left(1 - \frac{1}{n} \right) \frac{Y_{PS}}{6} \right] - Z_P \left[Y_P - \frac{Y_{PG}}{6} + (n - 1) \frac{Y_{PS}}{6} \right] \\
&+ \frac{V_{SM}}{I_S} \left[\frac{Y_{SG}}{2} + \left(1 - \frac{1}{n} \right) \frac{Y_{PS}}{2} \right] + \frac{V_{PM}}{I_P} \left[\frac{Y_{PG}}{2} - (n - 1) \frac{Y_{PS}}{2} \right]
\end{aligned} \tag{2.44}$$

where the currents are defined at the M terminals of the primary and secondary windings.

The difference between the currents entering and leaving the comparator may also be determined:

$$I_{PN} = I_{PM}(1 + \beta_P) \tag{2.45}$$

$$I_{SN} = I_{SM}(1 + \beta_S) \tag{2.46}$$

where

$$\beta_P = Z_P \left(\frac{Y_{PG}}{2} + \frac{Y_{PS}}{2} \right) + \frac{1}{n} Z_S \frac{Y_{PS}}{2} + \frac{V_{PM}}{I_P} (Y_{PG} + Y_{PS}) - \frac{1}{n} \frac{V_{SM}}{I_S} Y_{PS} \tag{2.47}$$

$$\beta_S = Z_S \left(\frac{Y_{SG}}{2} + \frac{Y_{PS}}{2} \right) + n Z_P \frac{Y_{PS}}{2} - \frac{V_{SM}}{I_S} (Y_{SG} + Y_{PS}) + n \frac{V_{PM}}{I_P} Y_{PS} \tag{2.48}$$

Three other defined errors, due to capacitive currents, are

$$\varepsilon_{C,NN} = \varepsilon_{C,MM} - \beta_P + \beta_S \tag{2.49}$$

$$\varepsilon_{C,MN} = \varepsilon_{C,MM} + \beta_S \tag{2.50}$$

$$\varepsilon_{C,NM} = \varepsilon_{C,MM} - \beta_P \tag{2.51}$$

The capacitive error of a current comparator as well as the capacitive error of current transfomers may be divided into two parts: that due to the internal voltage drops across the windings, and that due to externally imposed voltages. (For the latter phenomenon a voltage coefficient concept is usually used.[10]) It should be noted that the voltage appearing at marked terminals may be affected by the voltage drops in the leads. So, for convenience, the marked terminals are defined at the end of the leads and these are brought to ground potential. The terms V_{PM}/I_P and $-V_{SM}/I_S$ then become equal to the primary and secondary lead impedances, because $V_{PM} = Z_{PL} I_P$, $V_{SM} = -Z_{SL} I_S$. Of course, in eqn. (2.43), a term due to the capacitive current reaching the marked lead should be added.

For every uniformly wound toroidal primary and secondary winding placed in one or in several layers, an estimate can be made of the capacitive error. The required steps in the procedure are

1   Calculation or measurement of the winding voltages
2   Calculation or measurement of the distributed capacitances
3   Location of the capacitive current paths
4   Summation of the ampere-turns and the error estimation.

The results obtained may not be very accurate but even so valuable data for current comparator design are obtained, especially if the estimated error is less than 1 ppm.

As a typical example, the capacitive error is given as follows for a multi-section primary winding placed side by side in the same layer and a secondary

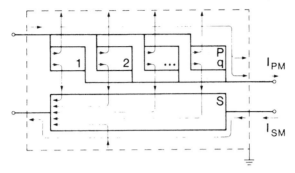

Fig. 2.12 *Capacitive ampere-turns in current comparator with single-layer multisection primary winding*

distributed in one layer also (Fig. 2.12):

$$\varepsilon_{C.MM} = Z_S \left\{ Y_S - \frac{Y_{SG}}{6} - \left[1 - \frac{3 - (1/q)}{2n} \right] \frac{Y_{PS}}{6} \right\}$$
$$- Z_P \left\{ Y_P - \frac{Y_{PG}}{6} + \left[\frac{3 - (1/q)}{2} n - 1 \right] \frac{Y_{PS}}{6} \right\}$$
$$+ \frac{V_{SM}}{I_S} \left[\frac{Y_{SG}}{2} + \left(1 - \frac{1}{n} \right) \frac{Y_{PS}}{2} \right] + \frac{V_{PM}}{I_P} \left[\frac{Y_{PG}}{2} + (1 - n) \frac{Y_{PS}}{2} \right] \quad (2.52)$$

where q is the number of primary sections and Z_P is the leakage impedance of all q sections in parallel.

2.5.2 Shield excitation

Reducing the error caused by internal capacitances when the ratio is very large is a difficult problem. The number of primary turns can be no fewer than unity; hence the number of secondary turns must be quite large, and the voltage drop across the secondary, which is attributable only to its internal impedance, will be correspondingly high.

Even more serious a problem is imposed when small currents are compared. Then the number of turns is bound to be large for sensitivity reasons, and the error is larger because the ratio between the capacitive currents and the winding currents is increased.

The magnetic shield, whose role in reducing the magnetic errors of the comparator has already been explained, can be used at the same time to reduce the capacitive error. The idea is to excite the shield so that the electromotive force induced in the secondary winding is of such magnitude and phase that it cancels the voltage drop in that winding.[4] The potential in the secondary winding then will be nearly constant and, if this is ground potential, no capacitive current will flow in the secondary winding – provided, of course, that it is shielded electrostatically or that the voltage in the primary winding is negligibly small.

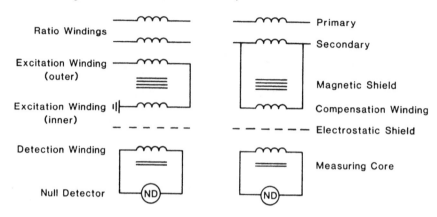

Fig. 2.13 *Excitation and compensation windings*

The excitation current necessary to produce the magnetic flux in the shield must not be imposed, however, on the measuring core. To realise this, two windings with the same number of turns are used, one wound outside and one inside the magnetic shield (Fig. 2.13). These two windings are then connected in series in such a manner that a magnetic field is produced in the shield, but no magnetising force is imposed on the measuring core. The excitation winding then provides a method of controlling the voltages across the secondary and primary windings, without influencing the comparison of their currents.

2.5.3 The compensation winding
The compensation winding inherently provides the means for maintaining the whole secondary winding at ground potential. Moreover, the need for an additional source of shield excitation and the means for its adjustment have vanished.

As has been explained, the excitation winding consists of two parts: one winding is wound between the measuring (ΣI sensing) core and the heavy magnetic shield, and the other is wound outside the shield. The latter winding, however, may be incorporated with the secondary (Fig. 2.13). The inside part of the excitation winding must now have the same number of turns as the secondary. Also, if this inside part of excitation winding is connected in parallel with the secondary, compensation of voltage drops all along the secondary winding is obtained inherently.[11] The secondary and inner part of the excitation winding, now renamed the compensation winding, form a composite ratio winding. The energy required to overcome the voltage drops in the secondary winding is obtained from the primary winding circuit.

At balance the primary ampere-turns are equal to the sum of ampere-turns formed by the currents in the secondary and the compensating winding. The magnetomotive force seen by the measuring core is the same for both paths. Most of the current carried by the composite winding will be, however, in the outside winding, that is the secondary. This is because the primary winding and secondary winding together with the magnetic shield form a current transformer. Only the error current of this current transformer will pass through the compensation winding. The voltage drop across the compensation winding is consequently very small. Inherently the shield is excited so that induced voltage per turn in the secondary winding is equal to the leakage impedance voltage drop per turn and the most desirable result is obtained. Everywhere in the secondary winding, which has a large number of turns, the potential is the same and no capacitive currents are present. It can be said that the voltage drop in the secondary winding has been compensated by exciting the shield.

In a shield-excited or compensated current comparator the whole secondary winding remains essentially at ground potential. This is achieved at the expense of an increased voltage drop in the primary or uncompensated ratio winding, but this can be tolerated since generally this winding has a lower number of turns and hence the capacitive current ampere-turns are small. To prevent the capacitive currents due to the distributed capacitance between the primary and secondary windings from reaching the secondary winding with the larger number of turns, it is advisable to include a thin electrostatic shield between the primary and secondary ratio windings. Capacitive currents resulting from the increased primary voltage drop caused by the induced voltage from the shield excitation will flow only through a few primary turns and the electrostatic shield. Consequently, the total capacitive error will be reduced.

The implementation of the compensation winding technique greatly increased the usefulness of the current comparator as a ratio standard in precision measurements.

2.6 Two-stage current transformers

If a toroidal compensation winding is uniformly distributed around the measuring core, if the primary and secondary winding leakage fluxes are prevented by the magnetic shield from reaching the measuring core, and if the flux in the measuring core is zero, the voltage drop in that compensation winding will be essentially resistive. The magnitude of that resistive voltage drop is very small because, first, only the transfomer (shield) magnetising current flows through it and, second, this winding can be made of large wire.

Since the compensation and detection windings occupy more or less the same location in the overall configuration, they can be combined. The null detector may be connected to the compensation winding and the detection winding omitted. A zero indication of the null detector will then mean that in the measuring core the 'working flux' is of such magnitude as to induce the voltage needed to cancel the resistive voltage drop in the compensation winding. Theoretically the sum of the primary and combined secondary ampere-turns is not zero, but in practice it is virtually zero because the measuring core magnetising

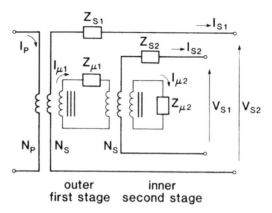

Fig. 2.14 *Equivalent circuit of a two-stage current transformer*

current is very small. The current comparator now behaves as a two-stage transformer with a short-circuited second stage.[11,12]

What are the features of such an improved two-stage current transformer? Around the inner toroidal magnetic core a second-stage secondary winding is uniformly wound with N_S turns. Then the hollow toroidal magnetic shield is assembled and a first-stage secondary winding with the same number of turns N_S is introduced. Finally, the primary winding with N_P turns is installed. Figure 2.14 shows the equivalent circuit of such a two-stage transformer. All magnetic and capacitive errors are neglected but, to account for the measuring core and the shield magnetising currents, fictitious loss-free windings of N_S turns are shown connected to impedances $Z_{\mu 1}$ and $Z_{\mu 2}$.

It should be emphasised that the rather simple equivalent circuit of Fig. 2.14 has been made under the assumption that the mutual impedance between the primary winding and the second or inner secondary winding (referred to the secondary) is equal to the mutual impedance of the first or outer secondary winding to the inner secondary winding:

$$\left(\frac{N_S}{N_P}\right)^2 Z_{P\ S2} = Z_{S1\ S2} \tag{2.53}$$

where $Z_{P\ S2}$ is the mutual impedance between the primary and inner secondary winding, and $Z_{S1\ S2}$ is the mutual impedance between the outer secondary and the inner secondary winding. This relation has been achieved by magnetic shielding, and it can be checked experimentally by determining whether the current ratio is affected by changing the position of the primary winding. One must be aware, for example, that when designing the two-stage transformer with two side-by-side toroidal magnetic cores, eqn. (2.53) may not be correct. In other words, the same precautions regarding the magnetic error have to be taken into account when constructing a two-stage transformer as were taken when constructing a current comparator. The same reasoning holds for the capacitive error of a two-stage transformer.

Starting from the equations for the first and second magnetic circuits (see Fig. 2.14)

$$N_P I_P = N_S(I_{S1} + I_{\mu 1}) \tag{2.54}$$

$$N_P I_P - N_S I_{S1} = N_S(I_{S2} + I_{\mu 2}) \tag{2.55}$$

and adding the voltage balance equations for the two secondary circuits

$$V_{S1} = Z_{\mu 1} I_{\mu 1} + Z_{\mu 2} I_{\mu 2} - Z_{S1} I_{S1} \tag{2.56}$$

$$V_{S2} = Z_{\mu 2} I_{\mu 2} - Z_{S2} I_{S2} \tag{2.57}$$

then if the higher terms are neglected

$$I_{S1} + I_{S2} = \frac{N_P I_P}{N_S}\left(1 - \frac{Z_{S1} Z_{S2}}{Z_{\mu 1} Z_{\mu 2}}\right) - V_{S1}\frac{Z_{S2}}{Z_{\mu 1} Z_{\mu 2}} - \frac{V_{S2}}{Z_{\mu 2}} \tag{2.58}$$

The error, defined as

$$\varepsilon = \frac{N_S(I_{S1} + I_{S2})}{N_P I_P} - 1 \tag{2.59}$$

is

$$\varepsilon = -\frac{Z_{S1} Z_{S2}}{Z_{\mu 1} Z_{\mu 2}} - \frac{V_{S1}}{I_S}\frac{Z_{S2}}{Z_{\mu 1} Z_{\mu 2}} - \frac{V_{S2}}{I_S}\frac{1}{Z_{\mu 2}} \tag{2.60}$$

where

V_{S1}, V_{S2} = output voltages of the first (outer) and the second (inner) secondary windings

$I_S = I_{S1} + I_{S2}$

$Z_{\mu 1}, Z_{\mu 2}$ = magnetising impedances of the cores (referred to the secondary)

Z_{S1} = leakage impedance of the outer secondary winding

Z_{S2} = leakage impedance of the inner secondary winding, mostly its resistance

If both secondary windings are short circuited, then

$$V_{S1} = V_{S2} = 0$$

and

$$\varepsilon = -\frac{Z_{S1} Z_{S2}}{Z_{\mu 1} Z_{\mu 2}} \tag{2.61}$$

If both secondary windings are connected to a common burden Z_B (Fig. 2.15a), then

$$\frac{V_{S1}}{I_S} = \frac{V_{S2}}{I_S} = Z_B$$

and

$$\varepsilon = -\frac{Z_{S1}Z_{S2}}{Z_{\mu 1}Z_{\mu 2}} - Z_B\left(\frac{Z_{S2}}{Z_{\mu 1}Z_{\mu 2}} + \frac{1}{Z_{\mu 2}}\right) \tag{2.62}$$

If the outer secondary winding is connected to a burden Z_B and the inner secondary winding is short circuited, then

$$\frac{V_{S1}}{I_S} = Z_B$$

$$V_{S2} = 0$$

and

$$\varepsilon = -\frac{(Z_{S1} + Z_B)Z_{S2}}{Z_{\mu 1}Z_{\mu 2}} \tag{2.63}$$

If each secondary winding is connected individually to separate burdens of

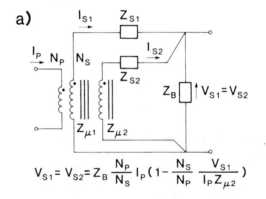

$$V_{S1} = V_{S2} = Z_B\frac{N_P}{N_S}I_P\left(1 - \frac{N_S}{N_P}\frac{V_{S1}}{I_P Z_{\mu 2}}\right)$$

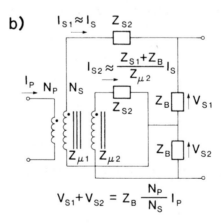

$$V_{S1} + V_{S2} = Z_B\frac{N_P}{N_S}I_P$$

Fig. 2.15 *Application of the two-stage current transformer*

value Z_B (Fig. 2.15b), then

$$\frac{V_{S1}}{I_S} = Z_B$$

$$\frac{V_{S2}}{I_S} = \frac{Z_B(Z_{S1} + Z_B)}{Z_{\mu 1}}$$

and

$$\varepsilon = -\frac{Z_B + Z_{S1}}{Z_{\mu 1}} \frac{Z_B + Z_{S2}}{Z_{\mu 2}} \tag{2.64}$$

Or, more generally for a transformer with n stages,

$$\varepsilon = -\prod_{i=1}^{n} \frac{Z_B + Z_{Si}}{Z_{\mu i}} \tag{2.65}$$

2.7 Null detectors

Any instrument connected to the network will disturb, to some extent, the state that existed before. When a load consisting of a null detector, a recorder or any other instrument is connected to the terminals of the detection winding, the sensitivity of the current comparator will be altered.

Figure 2.16 shows a simplified equivalent circuit of the current comparator and the instrument connected to its detection winding. The leakage reactance of the detection winding is neglected and its shunt capacitance transferred and joined with the load admittance. The detection winding magnetising inductance is

$$L = \frac{N_D^2}{R_m} = \frac{N_D^2}{\pi d} \mu S = N_D n \mu S \tag{2.66}$$

where

R_m = reluctance of the measuring core
N_D = number of turns in the detection winding
d = mean diameter of the toroid
μ = effective permeability of the magnetic core
S = cross-section of the magnetic core
n = turns density of the detection winding
ΣI = total ampere-turns passing through the toroidal opening exclusive of the detection winding ampere-turns
R = resistance of the detection winding
Y = input admittance of the connected instrument plus the equivalent detection winding shunt capacitance
V_D = voltage across the admittance Y

current comparator   null detector

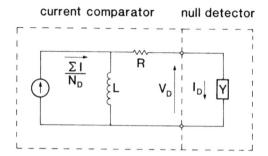

Fig. 2.16 *Equivalent circuit of the ampere-turn detector*

From Fig. 2.16 the Kirchhoff's voltage law gives

$$j\omega L \left(\frac{\Sigma_I}{N_D} - V_D Y \right) - R V_D Y = V_D \tag{2.67}$$

or

$$V_D = \frac{j\omega n\mu S}{1 + Y(R + j\omega L)} \Sigma I \tag{2.68}$$

If

$$Y = G + j\omega C$$

the voltage at the detection windings terminals is

$$V_D = \frac{j\omega n\mu S}{(1 + GR - \omega^2 LC) + j(\omega LG + \omega CR)} \Sigma I \tag{2.69}$$

and its amplitude is

$$|V_D| = k\omega n\mu S \left| \Sigma I \right| \tag{2.70}$$

where

$$k = 1/\sqrt{[(1 + GR - \omega^2 LC)^2 + (\omega LG + \omega CR)^2]} \tag{2.71}$$

The tuning factor k is maximum at $C = L/[(\omega L)^2 + R^2]$. Since $\omega L \gg R$ and $1/G \gg R$, this maximum becomes

$$k_{max} \simeq \frac{Q}{1 + DQ} \tag{2.72}$$

where

$$Q = \frac{\omega L}{R}$$

$$D = \frac{G}{\omega C}$$

Electronic null detectors have a very high input impedance – that is to say, a very low input admittance Y. On the other hand their sensitivity is large enough for the tuning of the detection winding to be unnecessary. The tuning factor (2.71) is therefore equal to one except where resonance occurs between the detection winding shunt capacitance and the winding inductance.

The above theoretical results indicate that the sensitivity of the current comparator, expressed in ohms as the ratio of the voltage at the detection windings terminals to the sum of ratio winding imbalance ampere-turns ΣI, may be reduced as well as increased by the presence of the input admittance of the null detector. However, owing to the high sensitivity of current comparators, the advantage of tuning is not usually realised if the ampere-turns to be compared are of the order of 1 ampere-turn or higher.

When, instead of voltage detection, the winding current is used as the indication of the current comparator balance, the instrument input admittance Y should be as large as possible. The detection winding current, measured by the attached instrument, is

$$I_D = \frac{j\omega n\mu S}{(1/Y) + (R + j\omega L)} \Sigma I \tag{2.73}$$

and obviously for $Y = \infty$ and $R = 0$ this current is equal to

$$I_D = \frac{\Sigma I}{N_D}$$

An important conclusion may now be stated. If the current measuring instrument is electronic its input impedance can be made very small, and such an instrument will directly measure the ratio winding imbalance. So, instead of balancing the comparator by providing a measured error current through the deviation winding, the current comparator may be allowed to balance itself naturally with the ratio winding error being measured directly by a current-sensitive instrument connected to the detection winding terminals.

The current comparator acts now as a two-stage current transformer with a short-circuited second stage. The number of turns of the detection winding may or may not be the same as the number of secondary turns, depending on the desired scale factor. The detection winding current may be resolved into in-phase and quadrature error components.

2.7.1 Vibration galvanometer

In high-precision alternating current bridges the vibrating galvanometer has been used extensively as a null indicating instrument. Current comparator balance for the fundamental frequency may be obtained using such an instrument.

The moving element of the vibrating galvanometer may be a small magnet suspended between the pole pieces of an electromagnet by a taut band or a similarly suspended moving coil in the air gap of a permanent magnet. The latter form of vibrating galvanometer is more common. Coarse tuning is accomplished by changing the active length of the suspension and fine tuning by adjusting the tension.

Although vibrating galvanometers can be constructed for operation at frequencies as high as 1000 Hz, their principal application is at the lower frequencies. At 60 Hz the sensitivity of a well designed vibration galvanometer may approach that of a good direct current galvanometer – that is, a few microvolts of imbalance may be readily detected. It should be noted however that the vibrating galvanometer is sharply turned and that, for example, a change of 0·5% in frequency from resonance will reduce the response to half value.

2.7.2 Electronic voltage null detectors

Owing to the relatively high sensitivity of the current comparator, any electronic AC millivoltmeter may be used as a null detector. Of course the noise of the instrument must be very low, preferably below $10\,\mu$V for measurements with a precision of 10^{-8}. To obtain this the connecting leads as well as the instrument itself should be electrostatically shielded.

An AC wave analyser may be successfully used as a null detector. When the null at the fundamental frequency is obtained the imbalance of higher frequencies may be measured by tuning the wave analyser accordingly.

2.7.3 Phase-sensitive detectors

In current comparators usually an in-phase and quadrature balance has to be achieved. Therefore it is desirable to use a phase-sensitive electronic amplifier as the null detector. Then the in-phase and the quadrature components may be adjusted almost independently.

It should be noted however that phase-sensitive detectors may be either tuned or gated. In the tuned detector the reference is a sine wave of fundamental frequency and the detector will be sensitive to that frequency only. In the gated detector, the reference is a square wave of fundamental frequency and the detector will respond not only to the fundamental frequency but also to any odd harmonics. Thus a false balance may be obtained where a fundamental component may be offset by a group of residual odd harmonics.

2.7.4 Direct current ampere-turn null detectors

The non-linear relationship between the magnetising force and the magnetic flux density for a ferromagnetic material affords a physical basis for the design of a direct current ampere-turn null detector.

The sensitive element of the direct current ampere-turn detector is a pair of toroidal magnetic cores situated inside a heavy magnetic shield. These two cores are magnetised in opposite directions to one another by an alternating modulation current. Owing to the shape of the hysteresis (B–H) loop, higher harmonics will be generated. When no direct current ampere-turns are present only odd harmonics will be generated because then the B–H loop is symmetrical on the plus and minus sides of the zero axis. If the direct current ampere-turns are present however the existing voltage will saturate the cores more easily in one direction that in the other. The waveform will not then be symmetrical about the zero axis and will contain even harmonics. A detection winding can be configured so that the fundamental and odd components of the induced voltage cancel one another while the second-harmonic components add. Other even harmonic voltages will also be present but these can be eliminated by tuning the detector. The net result of such an arrangement is a second-harmonic voltage whose amplitude is reasonably proportional to the direct current ampere-turns.[13]

In the first DC current comparator[14] the excitation generator had more or less a sinusoidal wave shape and the cores could be driven to the various flux levels. A separate pair of detection windings connected to the second-harmonic detector were used.

To improve the essential characteristics of the direct current ampere-turn detector, that is to reduce the noise, drift, and memory effects, a new approach to the modulation/detection circuit was adopted. Instead of a sinusoidal voltage a square wave voltage is used for core excitation. To obtain the square wave voltage the Royer oscillator[15] is used in which two transistors exchange states when the flux in an associated magnetic core approaches saturation level. This is achieved by a pair of auxiliary windings connected to transistor bases. It was discovered that a significant improvement in the direct current ampere-turn detection could be achieved if the sensing toroidal cores of the comparator not only act as detectors but also provide the switching signal for the transistors in the oscillator. This has an effect that flux detection cores always saturate by almost the same amount even if the temperature changes. The frequency of oscillation f is given by

$$f = \frac{V}{4NSB_{sat}} \tag{2.74}$$

where

V = supply voltage
N = number of turns on each sensing core
S = area of the sensing core
B_{sat} = saturation flux density

The modulator/detector circuit[16] consists of two high-permeability magnetic sensing cores upon each of which are windings that are connected in series to one another, a square wave oscillator connected to these windings via a pair of resistors and controlled by the sum of voltages induced in these windings, and a peak detector connected between the middle point of the linear (unsaturated) oscillator output transformer and the middle point of windings on the sensing cores (see Fig. 7.1). Although the core magnetising current is the same, owing to the core mismatch the voltage collapse in one core will occur earlier than in the other. As a consequence a peak voltage of odd harmonics will appear at the detector twice each cycle. When no direct current ampere-turns are present the positive and negative peaks will be the same, and the detector which measures the difference of their height will read zero. However, when direct current ampere-turns are present the saturation process in one corner of the hysteresis loop will not be the same as in the other, and even harmonics will be generated. As a consequence one peak, say the positive, will be higher than the negative, or vice versa, and the output of the peak detector will be proportional to the sum of all direct current ampere-turns linking the measuring cores.

The role of heavy magnetic shielding is essential for the performance of direct current ampere-turns detection. It prevents the leakage flux from reaching the measuring cores and disturbing the very delicate process of even harmonic generation. The magnetic shield may also be used for the suppression of the ripple components present in the direct current which are compared. To obtain this a short-circuited winding may be wound above the shield. However, if a large mutual coupling between the ratio windings is desired to ease the tracking of direct currents this winding should not be used.

The state of the art is that the sensitivity and drift stability of the direct current ampere-turns detector are of the order of a few microampere-turns.

2.8 Electronically aided current ratio devices

Throughout the whole development period of the current comparator and high-precision current measurements in general, some electronic circuitry has been employed. In fact, early efforts to implement the current comparator concept arose from attempts to apply control circuit techniques to high-precision measurements.[17,18] Feedback was used to bring the core magnetic flux in the current transformer, or in the current comparator, to zero.

Although electronic amplifiers were capable of providing single or multiple balances, workers in precision electrical measurements were reluctant to accept the new technique. There was a healthy disrespect for all active devices, including even null detectors. The drawbacks of such active circuits are

1 Susceptibility to breakdown or, even worse, to partial failure without a visible sign of failure having occurred
2 The inconvenience of having to supply power to the measuring system
3 The possibility of interaction (beat) between the power supply and the measuring circuit
4 The constructional difficulties of stabilising a high-gain feedback loop.

In spite of these problems, electronic circuitry has been and is going to be used more and more in the future. Automation of the measurement process is inevitable. Moreover, some high-precision measurements cannot be performed without the aid of electronic circuitry.

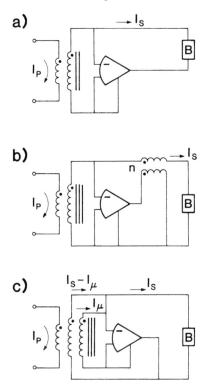

Fig. 2.17   *Electronically controlled current transformer circuits*

In Fig. 2.17 three different circuits are shown in which an electronic amplifier is used to aid the single-stage transformer. As is well known the error of the current transformer is due to the magnetising current, and this current is proportional to the induced voltage needed to establish the current in the burden. This voltage may be made considerably smaller if the burden is connected in the feedback loop of an electronic amplifier as shown in Fig. 2.17a. This amplifier will impose on that burden such a voltage as to remove almost all the voltage induced in the secondary winding. The voltage that will remain is approximately given by dividing the burden voltage by the amplifier gain. The transformer now actually performs as a measuring rather than as a power transmitting device. A small induced voltage is still necessary to cancel the internal burden of the secondary winding. However the magnetic flux in the core will be greatly reduced and the magnetising current will decrease. The current transformer error will be that which would be present if the burden was removed and the secondary winding short-circuited.

To match the amplifier output impedance with the burden, sometimes an additional transformer has to be used. Such a scheme is shown in Fig. 2.17b. The transformer will however introduce stability problems additional to those already existing, and these cannot be easily solved because the burden and the supply impedance, which are not under the designer's control, appear in the feedback loop. On the other hand the transformer separates the amplifier output from the burden and so the burden ground may be connected to the electronic power supply ground.

Complete removal of the internal burden of the secondary winding may be obtained if an additional detection winding is used as shown in Fig. 2.17c. Here again, the stability problems are to some extent increased because the feedback is closed through two windings of the transformer.

In all of the schemes shown in Fig. 2.17, and in others that are similar, the amplifier is rated for the burden power. The stability problems are solved by the implementation of dynamic compensation and by reducing the feedback loop gain for higher frequencies.

The precision of the current ratio measurement may be greatly improved if a two-stage transformer with concentric toroidal construction, introduced in the

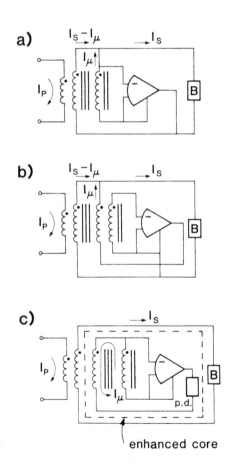

Fig. 2.18 *Electronically aided two-stage current transformer circuits*

current comparator for magnetic shielding, is combined with the electronic amplifier feedback technique.

Figure 2.18 shows a few possible circuit arrangements. In all of these the amplifier provides only a very small portion of the burden power. In Fig. 2.18a the voltage at the terminals of the second (inner) secondary winding is connected to the amplifier input. The amplifier drives the magnetising current through this winding so as to reduce the magnetic flux in the second (inner) core to a very small value. Actually the induced voltage of that core is equal to the product of the magnetising current of the first core and the leakage impedance of the second (inner) secondary winding.

To further reduce the magnetic flux in the inner core and to bring the magnetising current of that core almost to zero, a separate detection winding may be used as shown in Fig. 2.18b. The error of such an electrically aided two-stage transformer or current comparator may be well within 1 ppm.

Figure 2.18c shows an enhanced core denoted by dashed lines. A detection winding made of N turns of high-gauge copper wire is uniformly wound on the inner toroid core and electrostatically shielded. Then the magnetic shield or a first-stage transformer core made in the form of a hollow toroid is assembled and a winding of the same number of turns N is installed. This assembly is then enclosed in a copper box and the amplifier connected as shown.

This toroidally shaped copper box contains the enhanced core. The primary and secondary ratio windings may now be installed and an error-free transformer will be obtained. The feedback amplifier will always supply such a

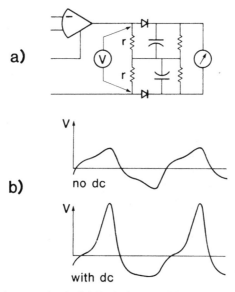

Fig. 2.19 *(a) Peak detector circuit (b) Typical magnetising current waveforms, with and without DC in I_P*

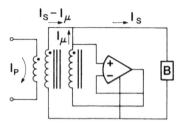

Fig. 2.20   *Two-stage transformer aided by voltage follower*

magnetising current as to establish the required burden voltage.[19] This magnetising current does not impose ampere-turns on the inner core because it passes through the same number of turns N in the opposite sense.

On the other hand because the induced voltage in the inner secondary winding is zero the secondary current must be exactly equal to the primary current multiplied by the turns ratio.

The capacitive error of the inner excitation (detection) winding is negligible because all of this winding is essentially at ground potential. The capacitive error of the outer excitation winding needs to be compensated only once for all possible configurations of the ratio windings and for all burdens. The only error that theoretically still remains is the capacitive error of the ratio windings, and this has to be evaluated or compensated.

Since the magnetising current of the outer magnetic core is available as a separate entity, it is possible to examine its waveform. Of particular importance is whether this waveform contains any even harmonics, as these will indicate the possible presence of a direct current component in the primary current I_P. Even harmonics can be detected using a peak detector circuit, installed for example at the point PD in Fig. 2.18c and having a form such as is shown in Fig. 2.19a. Typical waveforms of the magnetising current with and without a direct current component present in the primary current I_P are shown in Fig. 2.19b. Knowledge of the presence of such direct current components is extremely important in power electronic circuits.

It should be mentioned that many other schemes exist which give the same or very similar results.[20,21] The control feedback to bring the magnetic flux of the second (inner) core to zero may be established by using an amplifier follower as, for example, shown in Fig. 2.20.

A different approach for obtaining an error-free transformer is also possible. The second-stage current, which is very close to the magnetising current of the first stage, may be measured and a voltage proportional to this error obtained. Then a current proportional to this voltage may be injected into the burden.

References

1 W. Rogowski and W. Steinhaus. 'Die Messung der magnetischen Spannung. (Measurement of magnetic potential).' *Arch. Elektrotechn.*, **1**, 1912, pp. 141–50

2 H. S. Baker. 'Current transformer ratio and phase error by test ring method.' *AIEE Proceedings*, **36**, 1917, pp. 1173–83

3 N. L. Kusters and W. J. M. Moore. 'The current comparator and its application to the absolute calibration of current transformers.' *AIEE Trans. Power Apparatus and Systems*, **80**, April 1961, pp. 94–104

4 P. N. Miljanic, N. L. Kusters, and W. J. M. Moore. 'The development of the current comparator, a high-accuracy A-C ratio measuring device.' *AIEE Trans. Communications and Electronics*, **81**, November 1962, pp. 359–68

5 A. H. M. Arnold. 'Leakage phenomena in ring-type current transformers.' *J. IEE*, **74**, 1934, pp. 413–23

6 A. H. M. Arnold. 'Dielectric admittances in current transformers.' *Proc. IEE*, **97**, part 2, 1950, pp. 727–34

7 A. H. M. Arnold. 'The effect of capacitance on the design of toroidal current-transformers.' *Proc. IEE*, **97**, part 2, 1950, pp. 797–808

8 J. J. Hill and A. P. Miller. 'The design and performance of high-precision audio-frequency current transformers.' *Proc. IEE*, **108**, part B, May 1961, pp. 327–32

9 P. N. Miljanic. 'Capacitive error in current comparators.' *IEEE Trans. Instrumentation and Measurement*, **IM-13**(4), December 1964, pp. 210–16

10 N. L. Kusters and W. J. M. Moore. 'The effect of winding potentials on current transformer errors.' *IEEE Trans. Communications and Electronics*, **81**, July 1962, pp. 186–91

11 N. L. Kusters and W. J. M. Moore. 'The compensated current comparator: a new reference standard for current transformer calibrations in industry.' *IEEE Trans. Instrumentation and Measurement*, **IM-13**(2–3), June–September 1964, pp. 107–14

12 H. B. Brooks and F. C. Holtz. 'The two stage current transformer.' *AIEE Trans.*, **41**, June 1922, pp. 382–91

13 F. C. Williams and S. W. Noble. 'The fundamental limitations of the second-harmonic type of magnetic modulator as applied to the amplification of small d.c. signals.' *J. IEE*, **97**, part 2, August 1950, pp. 445–59

14 N. L. Kusters, W. J. M. Moore, and P. N. Miljanic. 'A current comparator for precision measurement of D-C ratios.' *Communications and Electronics*, **70**, January 1964, pp. 22–7

15 G. H. Royer. 'A switching transistor D-C to A-C converter having an output frequency proportional to the D-C input voltage.' *AIEE Trans. Communications and Electronics*, **74**, July 1955, pp. 322–4

16 M. P. MacMartin and N. L. Kusters. 'A direct-current-comparator ratio bridge for four-terminal resistance measurements.' *IEEE Trans. Instrumentation and Measurement*, **IM-15**(4), December 1966, pp. 212–20

17 P.N. Miljanic. 'Servo sistemi u elektricnim merenjima (Servo systems in electrical measurement).' Doctorate Thesis, Institut Nikola Tesla, Belgrade, 1957

18 I. Obradovic, P. Miljanic, and S. Spiridonovic. 'Prufung von Stromwandlern mittels eines Stromkomparators und eines elektrischen Hilfssystems (Testing of current transformers with a current comparator and an auxiliary electrical system).' *ETZ(A)*, **78**(19), October 1957, pp. 699–701

19 D. L. H. Gibbings. 'A circuit for reducing the exciting current of inductive devices.' *Proc. IEE*, part B, May 1961, pp. 339–43

20 O. Petersons. 'A self-balancing current comparator.' *IEEE Trans. Instrumentation and Measurement*, **IM-15**(1–2), pp. March–June 1966, pp. 62–71

21 T. M. Souders. 'Wide band two-stage current transformers of high accuracy.' *IEEE Trans. Instrumentation and Measurement*, **IM-21**(4), November 1972, pp. 340–5

Chapter 3

Design and construction

3.1 Introduction

The design of a current comparator involves many judgments and compromises. Current level, frequency range, sensitivity, accuracy, availability of material, ease of manufacture, and cost are all factors to be considered. In this chapter the way in which the design of the various components of a comparator may be affected by these factors is examined. To facilitate this discussion these components are identified in Fig. 3.1, which shows their location outward from the magnetic core. A cut-away view of an alternating current comparator is shown in Fig. 3.2, and the arrangement of the components for a direct current comparator in Fig. 3.3.

To detect short circuit and insulation faults arising during construction and to confirm the effectiveness of the shielding as it is installed, the probe test described in Section 3.9.1 should be carried out at frequent intervals during the construction.

3.2 Magnetic materials

Ferromagnetic materials are used in the current comparator for the ampere-turn sensing core and shielding. For high resolution of an ampere-turn balance condition, a material with high initial permeability is required. However, for shielding purposes a lower permeability and hence a less expensive material is usually adequate.

Magnetic materials commonly used for current comparators are

Supermalloy An 80% nickel-iron with an initial permeability greater than 40 000. This material is available in tape-wound toroidal cores and is particularly useful in low-level current comparator bridges where resolutions approaching 4 nanoampere-turns are sometimes desired.

Round HyMu 80 An 80% nickel-iron with an initial permeability greater than

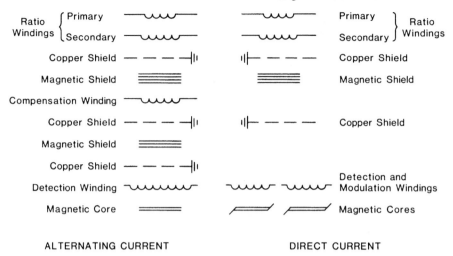

Ratio Windings	Primary				Primary	Ratio Windings
	Secondary				Secondary	
	Copper Shield				Copper Shield	
	Magnetic Shield				Magnetic Shield	
	Compensation Winding					
	Copper Shield				Copper Shield	
	Magnetic Shield					
	Copper Shield					
	Detection Winding				Detection and Modulation Windings	
	Magnetic Core				Magnetic Cores	

ALTERNATING CURRENT   DIRECT CURRENT

Fig. 3.1 *Components of a current comparator*

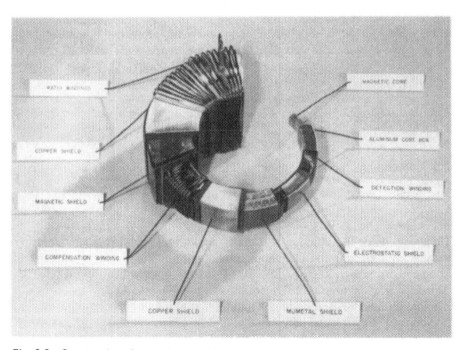

Fig. 3.2 *Construction of current comparator*

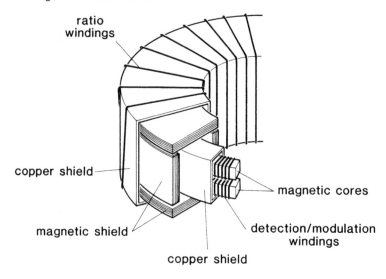

Fig. 3.3 *Arrangement of components of a direct current comparator*

20 000 and having a 'round' hysteresis loop (*B–H* curve). Tape-wound toroids of this material are used for the modulator cores of direct current comparators. 'Square' loop materials are to be avoided in this application because of their tendency to cause oscillations.

HyMu 80   An 80% nickel-iron with an initial permeability greater than 12 000, which is usually adequate for current comparators used at a higher current level such as in current transformer calibration. Available in tape-wound toroids or in sheet form for making flat shielding rings.

A 50% nickel-iron, grain-oriented, high-saturation-flux material   Tape-wound toroids are used for magnetic shielding, particularly where energy transfer is involved.

Grain-oriented silicon iron   Lower-cost tape-wound toroids used for magnetic shielding.

Mumetal   A 77% nickel-iron available in large sheets and suitable for making flat shielding rings and other shielding containers.

3.2.1 Annealing

Magnetic materials subjected to stresses and strains from the manufacturing process, or even from the installation of windings, suffer a deterioration in their magnetic properties. Lightly stressed items usually regain their original characteristics if allowed to return to their relaxed state. Others, such as shields whose fabrication has involved cutting, bending, and welding, must be annealed.

Annealing high-permeability magnetic materials is a complex process which involves heating to a temperature approaching 1200 C, maintaining that temperature for several hours and then cooling at various rates over a further period of several hours. To prevent sticking the material must be coated before-hand with aluminium oxide or similar material, and to avoid scaling and reduce contaminants the whole process is carried out in a hydrogen atmosphere. The specific temperature and heating/cooling programme is critical in determining the magnetic characteristics.

If the material has been overstressed or deformed after complete annealing, the magnetic characteristics can still be recovered to a large extent by what is known as 'stress relief annealing'. This involves heating and maintaining the material at a temperature of about 800°C for a period of two to three hours in a hydrogen atmosphere.

When these materials are used in the construction of a current comparator, proper care must be exercised to avoid residual stresses being imposed. Usually this is provided by enclosing the structure in a heavy gauge copper container.

3.2.2 Magnetic core testing

It is good practice to check the magnetic characteristics of cores and shields before they are actually used in the construction of a current comparator. Later it may be necessary to determine whether the material has retained any direct current magnetisation that may have resulted from winding continuity tests or switching transients. Although the magnetically shielded current comparator, unlike the current transformer, does not suffer a deterioration in accuracy due to direct current magnetisation, it does lose some of its sensitivity to ampere-turn imbalance.

A simple circuit for evaluating the condition of a magnetic core or shield, and removing any direct current magnetisation, is shown in Fig. 3.4. The values of R and C are selected to provide integration of the applied voltage at the frequency of the test, and r is small enough to avoid constraining the core magnetisation. The voltage at terminal B is then proportional to the flux in the core,

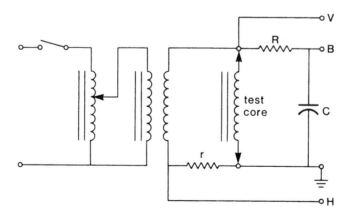

Fig. 3.4 *Circuit for testing magnetic cores*

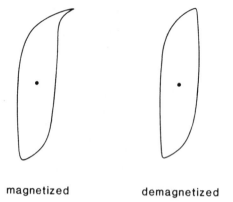

magnetized   demagnetized

Fig. 3.5 *Hysteresis curves of magnetised and demagnetised magnetic core*

while that at terminal H gives an indication of the magnetising and loss current. If these terminals are connected to the vertical and horizontal inputs of an oscilloscope, the hysteresis loop may be displayed. The voltage at terminal V enables calculation of the flux density without having to account for the attenuation of the integrating network.

Direct current magnetisation of the magnetic core is indicated by an asymmetrical hysteresis loop. To remove this magnetisation it is necessary to raise the voltage so as to drive the flux well into saturation, then reduce it slowly to zero. Care must be taken not to switch off the voltage before its zero value is reached, or the resulting transient may remagnetise the core. Also, if windings with large numbers of turns are present, it may be necessary to subdivide them or use a lower frequency to avoid voltage breakdown. Typical hysteresis loops showing the difference between a magnetised and an unmagnetised core are shown in Fig. 3.5.

A quick method for measuring the initial permeability of the magnetic core is to install a winding of two-conductor 'lamp' cord and apply a voltage to the ends of one of the conductors through a 1000 or 10 000 ohm resistor. The permeability can then be determined from the applied voltage, the value of the resistor, the voltage at the terminals of the other conductor, and the core dimensions.

3.3 Insulating materials

Insulating materials used in the construction of current comparators may be divided into three categories – tape, fibre and rigid. Tape insulation is preferably of the polyethylene or polytetrafluoroethylene (PTFE or Teflon) variety with a low dielectric constant ($\approx 2 \cdot 1$). Fibre insulation (also known as 'fish paper') is a vulcanised cellulose material available in sheet form of various thicknesses. It is mechanically tough yet, depending on thickness, relatively flexible and easy to work with. It has a dielectric constant varying from 4 to 5. For greater separations and where practicable, rigid insulations of the acrylic or phenolic resin type are used.

3.4 The ampere-turn balance detector for alternating currents

The ampere-turn balance detector for alternating current comparators consists of the magnetic core, the detection winding, an electrostatic screen, a magnetic shield, and a rigid copper box, all of which are toroidally configured.

3.4.1 The magnetic core
The magnetic core is usually supplied in an aluminium or rigid plastic box. The inner diameter is determined from the space required for the various windings and this usually requires that several trial designs be made. The cross-section depends on sensitivity requirements or, if the comparator is to be used in a cascade arrangement, the magnetising impedance required to support the load or burden imposed by the succeeding stage. Typical cross-sections vary from 1.25×0.6 cm to 2.5×2.5 cm. Usually 0.01 cm thick magnetic tape is adequate. Thinner material, while providing a somewhat lower loss component, also exacts a fill-factor penalty. The effect of tape thickness on various parameters for a typical 80% nickel-iron material with an initial permeability of 70 000 is given in Table 3.1. These calculations were made for a $15.25 \times 10.15 \times 2.54$ cm Supermalloy core using a 2000-turn winding.

3.4.2 The detection winding
The detection winding is uniformly distributed around the toroid, usually and preferably in a single layer. Where more than one layer is required, some form of separation such as one or two layers of polyethylene or polytetra-fluoroethylene tape (dielectric constant ≈ 2) is desirable. The 'air turn' formed by the traverse of the detection winding, as shown in Fig. 1.1, can be eliminated by reversing and increasing the pitch so as to bring the end of the winding back around the toroid using only a few turns.

Table 3.1 *Effect of tape thickness at 0.01 tesla, 60 Hz*

Thickness, mm	Reactance ωL, kΩ	Resistance R_P, kΩ	Q $R_P/\omega L$	Tuned V/ampere-turn
0·1	193	910	4·7	450
0·05	182	1460	8·0	770
0·025	161	2820	17·3	1390

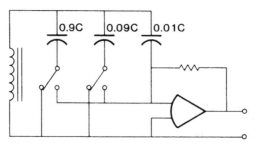

Fig. 3.6 *Variable-sensitivity error-sensing circuit for detection winding*

The number of turns in the detection winding depends on whether the ampere-turn balance is to be detected by a voltage null or a current null. Voltage sensing usually implies a high number of turns and a consequent large shunt capacitance, leading to a low self-resonant frequency and a high output impedance. Current sensing uses an electronic current-to-voltage interface which essentially applies a short-circuit across the detection winding. The output varies with the inverse of the number of turns and has the further advantage of being in phase with the ampere-turn imbalance. A problem associated with current sensing however is the maintenance of the direct current stability of the current-to-voltage interface, owing to the very high amplification of the amplifier input offset as given by the ratio of the feedback resistance to the resistance of the detection winding. This usually requires a large capacitor or at least some resistance in the detection winding circuit, which impairs the short-circuit feature. Alternatively a composite consisting of a high slew-rate amplifier for dynamic response and a slower low-offset amplifier for high direct current stability can be used.

If the current comparator is to be used at only one frequency, the detection winding can be tuned with a capacitance C (provided that it is not self-resonant at a lower frequency). Such an arrangement however offers a very high source impedance to a voltage sensor and, as shown in Fig. 3.6, current sensing with its low alternating current input impedance and direct current blocking can be used to advantage.

3.4.3 Electrostatic shield
The electrostatic shield protects the magnetic core and detection winding from ambient electric fields. It is constructed from thin copper foil with slits cut into the edges to enable wrapping around the detection winding. Overlapping areas of the shield must be insulated from each other, preferably with a rugged insulation such as vulcanised fibre (fish paper), to prevent a short-circuit being formed around the core. One or more layers of low-dielectric-constant tape insulation should be used to separate the shield from the winding, thus reducing capacitive coupling.

The two leads to the detection winding are brought out through a braided

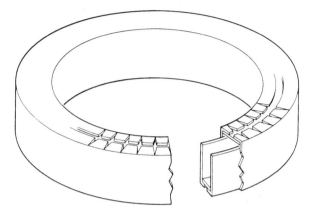

Fig. 3.7 *Construction of inner magnetic shield*

shield cable connected to the electrostatic shield. This shield is grounded only at the electronic interface. It is not connected to any other shield within the comparator itself, thus avoiding interaction with currents in other ground circuits.

3.4.4 Inner magnetic shield

The inner magnetic shield protects the magnetic core by intercepting the leakage fluxes of the compensation winding. It is usually configured as two U-shaped toroidal troughs which nest together as shown in Fig. 3.7. It can be constructed by spot-welding Mumetal sheet sections together. Appropriately shaped sections of fibre insulation are used to prevent the two halves forming a short-circuit around the core.

3.4.5 Copper shield

To protect the inner magnetic shield from mechanical stress and to add additional electromagnetic shielding at the higher frequencies, a thick (1·25 mm) copper toroidal shield is added. To prevent a short circuit, an open slit is usually left at one of the inner edges of the toroid.

3.5 The ampere-turn balance detector for direct currents

The ampere-turn balance detector for direct currents consists of two magnetic cores, the detection/modulation windings, and an electrostatic screen.

3.5.1 The magnetic cores

For good symmetry and balance, the magnetic cores should be a matched pair, preferably manufactured at the same time from the same raw material. Magnetic material that has been annealed to produce a 'round' characteristic is essential. High squareness in the hysteresis characteristic, although perhaps providing greater sensitivity, should be avoided because of a tendency to induce oscillations. Typical cores are made from 0·05 mm tape to reduce the loss at the modulation frequency (≈ 700 Hz), have a cross-section of 0·635 × 1·27 cm, and are encapsulated in plastic or, for larger diameters, in an aluminium box. Core diameters should be kept as small as possible to minimise magnetic noise effects.

3.5.2 The detection/modulation windings

The modulation/detection function is realised with one single-layer uniformly distributed winding on each core. Both windings should be wound with the same pitch and have the same number of turns. The number of turns is a direct function of the sensitivity and the modulation voltage, so a compromise is necessary to keep the voltage within reasonable limits. For a modulation frequency of about 700 Hz, windings of about 1000 turns are typical.

3.5.3 The electrostatic shield

The two cores are enclosed as a pair in an electrostatic shield made from thin copper foil, suitably cut to enable a close fit. Overlapping edges are insulated from one another with fibre. The four leads to the windings are brought out through the shield via an insulated flexible shield which connects the electrostatic shield to an external ground.

3.6 The compensation winding

The compensation winding is used in alternating current comparators only. Since it shares current with one of the ratio windings, the two must have exactly the same number of turns so that variations in the current split will not affect the overall ampere-turn balance. The winding should be uniformly distributed around the toroid to avoid imposing leakage fluxes on the ampere-turn balance detector and be wound with the largest cross-section wire that is feasible to keep the winding impedance low.

For some applications, particularly current transformer calibration as discussed in Section 4.2, a second 'compensation' winding with the same number of turns as the other ratio winding is useful to provide a means for grounding that takes into account the ground current.

3.7 The magnetic shield

The design of the main magnetic shield is governed by two factors:

1   Protection of the ampere-turn balance detector from the effects of the leakage fluxes of the ratio windings and of other ambient magnetic fields
2   The requirements for energy transfer between the ratio windings (this applies only to alternating current comparators).

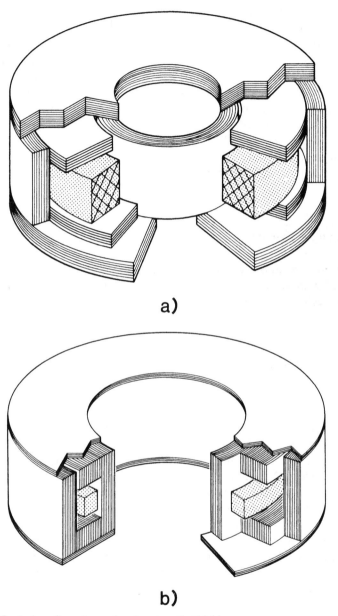

a)

b)

Fig. 3.8 *Typical configurations of main magnetic shield*

The protection requirement can usually be met by a relatively small amount of magnetic shielding. In critical applications where space is limited, it is made with Mumetal toroids and flat rings. Two rings sizes are used – one which fits within the toroids, and another which overlaps. A typical configuration is shown in Fig. 3.8a. The required thickness of the shielding may be determined by measuring its effectiveness using the probe test (Section 3.9.1). A reduction of 1000 or more in the probe's influence on the ampere-turn balance detector with and without the shield should be realised.

When energy transfer is involved, the cross-sectional area of the shielding is determined by the voltage per turn that must be generated without saturating the magnetic material. To permit higher flux densities, grain-oriented 50% nickel-iron or 3% silicon-iron alloys can be used. To minimise costs, four tape-wound toroids with one or two flat sheets of Mumetal on each side may be used, as shown in Fig. 3.8.

Fibre insulation must of course be incorporated in the assembly of the magnetic shield to prevent the formation of a short-circuit.

3.8 The outer copper shield

A rigid copper shield is used to enclose and to protect magnetic shields, such as those made with Mumetal and other materials with very high initial permeability, from mechanical stress.

3.9 Tests for shielding effectiveness

3.9.1 The probe test

The probe test is used to test the effectiveness of the shielding against winding leakage fluxes. The probe consists of a long narrow loop of several turns of wire which can be inserted through the toroid window and rotated axially with respect to the toroid itself. Since the current in the probe does not actually link with the core, it simulates a leakage flux. To prevent capacitive coupling at the higher frequencies, the probe is enclosed in a thin copper shield which is connected to ground. A typical apparatus is shown in Fig. 3.9.

A normalised measure of the effect of the probe can be obtained by dividing the voltage that it develops in the detection winding by the sensitivity of the detection winding to linked ampere-turns, with the probe rotated to its angle of maximum effect:

$$\text{normalised probe effect} = \frac{\text{maximum volts/probe ampere-turn}}{\text{sensitivity-volts/linked ampere-turn}}$$

Significant reductions in the probe effect should be observed as more and more shielding is added, as shown in Fig. 3.10. A more detailed study of this effect, including phase information, can be obtained using the circuit in Fig. 3.11.

Fig. 3.9 *Probe test apparatus*

3.9.2 Radial and axial fields

Tests of the effectiveness of the shielding against radial and axial magnetic fields may be made using apparatus similar to that of Fig. 3.12. If the two coils are excited so that these fields oppose one another, a radial field emanating outward from the axis is imposed on the test core. If the excitation is such that the fields of the two cores aid one another, a field parallel to the axis is produced.

3.9.3 Rated current testing

A final, more realistic and quantitative check on the shielding effectiveness of a completed comparator can be made by installing two loosely fitting concentrated windings with the same number of turns and connecting them in series

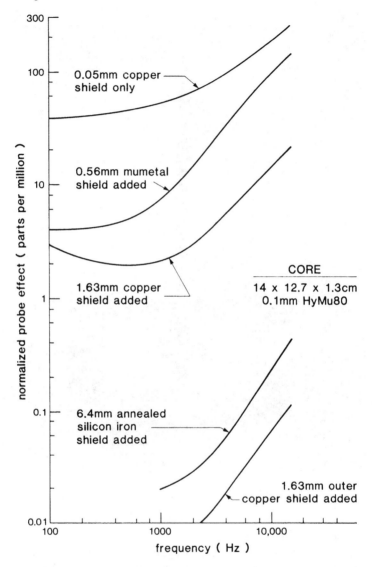

Fig. 3.10 *Shielding effectiveness: probe test*

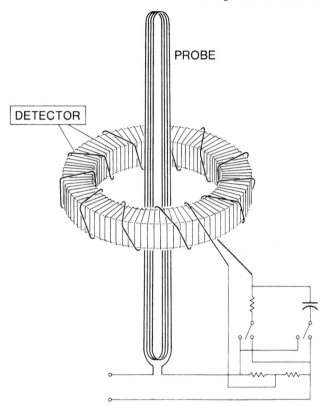

Fig. 3.11 *Probe test circuit*

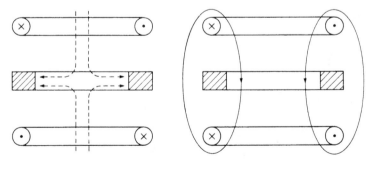

radial field test   axial field test

Fig. 3.12 *Axial and radial field tests*

opposition. With the current adjusted so that rated ampere-turns are imposed, the voltage of the detection winding is measured for various orientations of the two windings with respect to the comparator and one another, from which the normalised error can be determined.

3.10 The ratio windings

The ratio windings may be configured in either series/parallel or tapped winding format. Series/parallel windings make the most efficient use of the winding space but are limited in the number of ratios that can be provided. Tapped windings are more flexible and are essential where decade windings are required. Since their internal structure is fixed, they are also more easily corrected for capacitance error.

The ratio windings are usually installed with the lower current turns innermost. For some applications it may be desirable to separate the primary and secondary windings with a rigid copper shield.

3.10.1 Decade windings

Windings used for high-end decade switching in bridge applications can be easily wound by bunching ten conductors together and winding them as a

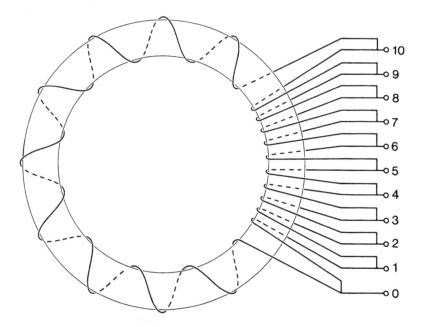

Fig. 3.13 *Winding configuration for a decade of single turns*

group. The conductors are then connected in series with a tap brought out to the switch from each connection point.

At the low end where ten single turns are required, a uniform distribution around the toroid is not readily attainable. Although the action of the magnetic shield tends to make this unnecessary, there are some techniques that can be used to spread the influence of a single turn. Figure 3.13 shows one such technique where the concentrated single turns are connected in parallel with ten distributed turns. The distribution of current between the two sets of turns however is determined by their relative leakage impedances.

3.10.2 Low-turn windings
Windings which do not have a sufficient number of turns to traverse the whole toroid may be distributed by using several equally spaced identical windings which are connected in parallel using a large cross-section ring bus. Another technique is to wind say ten turns evenly distributed around the toroid, then to reverse and wind a lesser number of turns back.

Chapter 4

Current transformer calibration

4.1 The compensated current comparator

The practical application of the current comparator to current transformer calibration is accomplished with a configuration that has been called the compensated current comparator.[1] The principal feature of this configuration is a winding located inside the main magnetic shield which has the same number of turns as one of the ratio windings located outside this shield. By connecting these two windings in parallel it is possible to operate the two ratio windings and the magnetic shield as a normal current transformer, thus enabling energy transfer between the primary and secondary circuits, without altering the overall ampere-turn balance imposed on the magnetic core.

The operation of the compensated current comparator may be explained by considering Fig. 4.1, where it is shown connected to a current transformer which, for this discussion, is considered to have no error. That is, the current ratio is exactly equal to the inverse of the turns ratio in both magnitude and phase, independent of its burden. Now the purpose of compensation is to enable the current comparator to supply, by means of its two ratio windings and shield excitation, the burden B_C while the current transformer supplies only its own burden B_T. This condition is realised if the voltage between points M and N is zero. One way to achieve this is to connect these two points together through a low-impedance link. However, the current in this connection must be taken into account so the compensation winding is included as part of the link. The compensation winding has the same number of turns as the secondary with which it is paralleled, and hence the net ampere-turns imposed on the magnetic core is unchanged. Being located inside the magnetic shield, it is unaffected by the shield excitation, it has a low resistance by design, and it has no voltage induced in it since the current transformer being calibrated is assumed to have no error and hence the flux in the magnetic core is zero. The link current itself is also very small, being equivalent to that required to magnetise the magnetic shield. Hence, for most practical purposes, the voltage between points M and N can be considered to be zero.

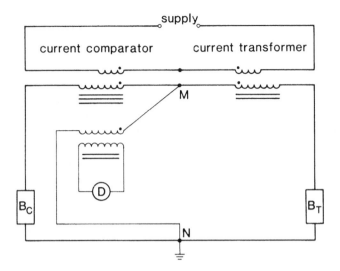

Fig. 4.1 *Elementary current transformer calibration circuit*

If the current transformer is not error-free it must be made to appear so to the current comparator by injecting an additional current into point M. This current is obtained from and measured by ratio error sets, which are described in Section 4.3.

An important feature of the compensated current comparator is that its burden B_C can be negative – that is, an energy source. Thus the ratio windings combined with the magnetic shield can be made to act as a supply transformer for the primary circuit. The usual high-current supply transformer can therefore be eliminated and replaced by a short-circuit. Except for the possibility of increased capacitance error caused by higher turn-to-turn voltage in the ratio windings, which can be adjusted for, the ratio accuracy of the current comparator itself is unaffected.

4.1.1 Design considerations

Current comparators for current transformer calibration are usually designed to support many ratios. Ratio variations may be realised with the primary winding either by series/parallel arrangements of individual winding segments, each with the same number of turns, by taps on a single winding, or by combinations thereof. Series/parallel arrangements have the advantage of automatically adjusting the current rating and of providing increased shielding against external magnetic fields as more and more sections are paralleled, but the ratio variations are somewhat limited and changing ratios is more tedious. Tapped windings provide greater flexibility in ratio selection and are easier to use, but the winding conductor sizes must be properly chosen to meet the current ratings required for each ratio.

Ratio variations may also involve changes in the secondary winding. Any such facilities for change in the secondary winding should preferably be duplicated in the compensation winding. (Alternatively an auxiliary external ratio transformer can be used.)

The total number of turns is determined by the maximum desired ratio and the burden the winding must support in conjunction with the magnetic shield. Enlarging the cross-section area of the shield is generally preferable to increasing the number of turns as capacitance error varies with the square of the number of turns. To avoid the capacitance problems associated with the high number of turns required by large ratios, current comparators can be connected in cascade (with all but one operating in the two-stage current transformer mode).

The dimensions of the magnetic core in current comparators for non-cascade applications are determined by sensitivity requirements. Where the comparator may be used as one component of a cascade arrangement, the core size must be such as to provide an adequate magnetising impedance in the compensation winding to support the burden imposed by the compensation winding of the succeeding comparator.

The minimum size of the magnetic shield is determined by the various tests

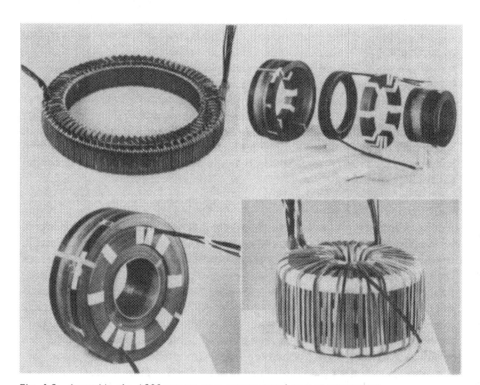

Fig. 4.2 *Assembly of a 1200 ampere-turn compensated current comparator*

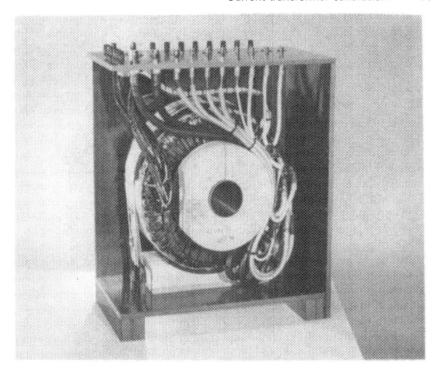

Fig. 4.3 *Assembled 2000 ampere-turn compensated current comparator*

for shielding effectiveness. Usually a thickness of 5 mm is adequate for this purpose. However, if the current comparator is to act also as the primary circuit supply transformer, the energy transfer requirements will govern and a more substantial cross-section will be needed.

Steps in the assembly of a 1200 ampere-turn tapped winding, compensated current comparator are shown in Fig. 4.2, starting with the compensation winding installed, adding the magnetic shield, and finally with the ratio windings in place. A view of an assembled 2000 ampere-turn comparator is shown in Fig. 4.3. The latter comparator features a 120/160/200 turn secondary winding and a 10/20/30/40/50/60/80/100/200 turn primary. A feed-through hole is provided for windings of less than 10 turns. Two compensation windings with the same taps as those of the primary and secondary windings are located inside the magnetic shield so that a secondary winding of 400 turns can be obtained by connecting the installed windings in series and using feed-through turns for the primary. The windings are designed for 100% overcurrent and the magnetic shield has the capability of transferring up to 8 kVA (2 volts/turn).

An exposed view of an audio-frequency current comparator for use at frequencies up to 16 kHz is shown in Fig. 4.4. These comparators feature series/parallel ratio variation with primary windings of 12 × 3 and 4 × 5 turns.

Fig. 4.4 *Audio-frequency current comparator*

4.1.2 Reduction of capacitance error

Capacitance errors are caused by turn-to-turn or shunt capacitances within a winding, by capacitances between windings, and capacitances from winding to ground. Shunt capacitance currents bypass all or part of a winding but remain within the circuit. Ground capacitance currents however pass through all or part of a winding and do not remain within the circuit. The direct approach to the reduction of the capacitance error is to keep the capacitances small by using the minimum practical number of turns and to make the voltages across the capacitances as low as possible. For designs where this is not feasible, the fact that the effects of shunt capacitance and ground capacitance oppose one another can be used to advantage.

The shunt capacitance error of current comparators which feature tapped windings for ratio variation can be reduced using the technique shown in Fig. 4.5. In this circuit the marked terminals of the ratio winding and its corresponding compensation winding are connected together at point M and a voltage is applied between the other ends of the two windings. Essentially all of the voltage drop occurs across the ratio winding which magnetises the magnetic shield. In the absence of any capacitance leakage, the current in the ratio winding would return through the compensation winding, thus imposing zero net ampere-turns on the magnetic core. However, with turn-to-turn shunt

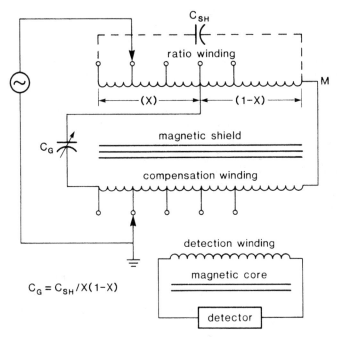

Fig. 4.5 *Adjustment of capacitive error correction: tapped windings*

capacitance present (represented in the diagram by an equivalent capacitance C_{SH}), less current passes through the ratio winding than in the compensation winding and the magnetic core becomes magnetised. To overcome this a capacitor C_G is connected from some convenient tap midway down the ratio winding to the end of the compensation winding and adjusted in value to achieve ampere-turn balance. This correction is independent of ratio since any loss of

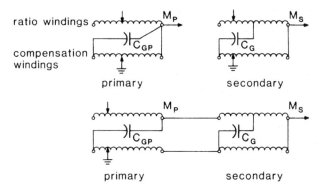

Fig. 4.6 *Method for adjusting capacitive error correction in configurations where the primary and secondary windings may be connected in series*

turns due to a tap change in one winding is recovered by a gain of turns in the other winding resulting from the required corresponding tap change.

In current comparators which provide for the primary winding to be connected in series with the secondary winding to obtain very large ratios, an additional correcting capacitor is required. The technique is illustrated in Fig. 4.6, which shows both the individual and combined winding arrangements. Note that when the primary winding is used as a true primary, that is alone, the voltage across the capacitor C_{GP} is negligible and hence it has no effect. Capacitor C_G is adjusted as previously described. Capacitor C_{GP} is adjusted when the windings are combined using the same technique.

Reduction of the capacitance error in current comparators where the ratio variation is achieved by series/parallel connections is less straightforward. Nevertheless, some adjustment can usually be realised by experiment.

4.1.3 Calibration of the basic compensated current comparator

The absolute calibration of compensated current comparators can be achieved by a step-up process starting with a self-calibration at 1/1 ratio. To accomplish this a set of current comparators is required with ratios that can be realised either by incrementing the lower ratio by one, or by the addition or multiplication of two lower ratios. Special corrections must be applied wherever a comparator is used in a circuit under non-standard conditions.

The calibration circuits and procedures described here were used to calibrate a set of current comparators with ratios ranging from 1/1 to 12/1 at frequencies up to 16 kHz.[2] At the higher frequencies great care must be taken in the shield arrangements and in ground connections. All connections to the various windings are made at a central point or junction box using twisted-pair leads that are shielded and insulated from one another. At power frequencies some relaxation, particularly in the shielding details, can of course be made.

A completely shielded current comparator, together with associated leads, junction box, and measuring circuit, is shown in Fig. 4.7. It is to be noted that the tuned null indicator or detector and the detection winding circuit are independently shielded and grounded. All other shields are connected to ground through the shield of the compensation winding lead to the grounded junction box. The detector D1 monitors the voltage of the point M_P which is brought to ground potential using a Wagner earth circuit (not shown). Similarly, detector D2 monitors the point M_S. Its potential is controlled by the adjustable voltage source which induces a voltage in the compensation winding lead through a small toroidal transformer with a single-turn secondary winding. This induced voltage is made equal and opposite to the voltage drop caused by the current in the compensation winding circuit, thus bringing the point M_S to ground potential. The network containing the components G, C and r is used to measure the proportional difference between the secondary currents of the reference comparator and the comparator under test, according to the relation

$$\delta = (G + j\omega C)r$$

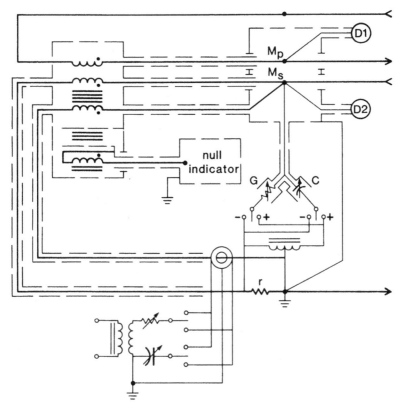

Fig. 4.7 *Arrangement of shields for using audio-frequency current comparators*

In the location in which it is shown in Fig. 4.7, the resistor *r* imposes a burden
on the secondary winding of the current comparator. Alternatively, this resistor
could have been located to the right of the ground point where it would not
have imposed such a burden. Since there is a difference in the errors of a current
comparator under 'zero burden' and 'with burden' conditions, a distinction
must be made. Methods for determining this difference will be evident in the

Fig. 4.8 *Construction of junction box used for audio-frequencies*

calibration circuits to be described, and care must be exercised in using the proper quantity in each step of the build-up process.

The junction box is shown in Fig. 4.8. The lower deck provides three independent terminals for the measuring points such as M_P and M_S, each of which is individually accessed by three coaxial connectors. The upper deck provides the terminals for making the various other connections to the comparator and measuring network. It includes a centre-tapped 0·2 ohm resistor, either side of which can be used in conjunction with the autotransformer as the resistor r.

4.1.3.1 The self-calibration circuit

The self-calibration circuit for 1/1 ratio comparators is shown schematically in Fig. 4.9. In this circuit the primary current is compared directly with the 'secondary' current and the proportional error is determined by adjusting the measuring network for comparator balance. (The secondary current here and in the other calibration circuits discussed in Section 4.1.3 is considered to be the sum of the secondary and compensation winding currents, defined at the point M. Also, the measuring circuit is shown in symbolic form only; for details see Fig. 4.7.) For this measurement to be valid, the point M must be at ground potential. This can be readily achieved by first connecting or short-circuiting point M to ground and balancing the comparator C by inducing a voltage in the compensation winding lead through core K. The short-circuit is then removed and the comparator rebalanced with the measuring network. The error of the current comparator is given by

$$\varepsilon_C = \delta$$

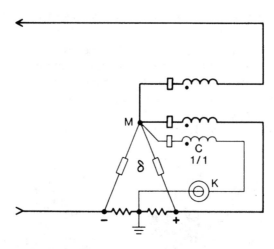

Self-calibration circuit

$$\epsilon_c = \delta$$

Fig. 4.9 *1/1 ratio self-calibration circuit*

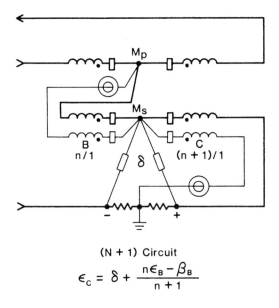

(N + 1) Circuit

$$\epsilon_c = \delta + \frac{n\epsilon_B - \beta_B}{n+1}$$

Fig. 4.10 *(n + 1)/1 ratio calibration circuit*

4.1.3.2. *The (n + 1) circuit*

Once a comparator has been calibrated at one ratio, the error of a second comparator of the next higher ratio can be obtained using the $(n + 1)$ circuit shown in Fig. 4.10. In this circuit, the primary current of comparator C is composed of the sum of the primary and secondary currents of comparator B. Two current sources are required which are capable of being adjusted relative to one another to achieve balance in one of the comparators. Points M_P and M_S are brought to ground potential by connecting both points to ground and adjusting the two compensation winding voltage sources K to balance their respective comparators. Final balance, with the ground connections to points M_P and M_S removed, is obtained with the measuring network.

A problem with this circuit and also the $(m + n)$ circuit to follow is that, owing to capacitive leakage, the secondary current I_{SB} of comparator B, as measured at the point M_S, is not the same as the current I'_{SB} at the other end of the windings at point M_P. A correction β must therefore be applied. The value of β is obtained with the β circuit to be described.

The balance equation for the $(n + 1)$ circuit is derived from the following relationships:

$$I_{SC} = \frac{I_{PC}}{n+1}(1 + \varepsilon_C)$$

$$I_{SB} = \frac{I_{PB}}{n}(1 + \varepsilon_B)$$

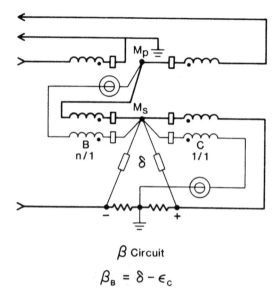

β Circuit

$$\beta_B = \delta - \epsilon_c$$

Fig. 4.11 *β coefficient measurement circuit*

$$I_{SC} = I_{SB}(1 + \delta)$$

$$I'_{SB} = I_{SB}(1 + \beta_B)$$

$$I_{PC} = I_{PB} + I'_{SB}$$

from which

$$\epsilon_C = \delta + \frac{n\epsilon_B - \beta_B}{n + 1}$$

4.1.3.3 The β circuit

The circuit for measuring the quantity β is shown in Fig. 4.11. It is similar to the $(n + 1)$ circuit except that the primary winding of comparator B is connected to ground rather than the point M_P, and comparator C has a 1/1 ratio only. The procedure for achieving balance is similar to that of the $(n + 1)$ circuit.

The balance equation for the β circuit is derived from the following relationships:

$$I_{SC} = I_{PC}(1 + \epsilon_C)$$

$$I_{SC} = I_{SB}(1 + \delta)$$

$$I'_{SB} = I_{SB}(1 + \beta_B)$$

$$I'_{SB} = I_{PC}$$

from which

$$\beta_B = \delta - \epsilon_C$$

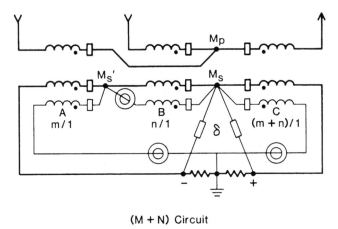

(M + N) Circuit

$$\epsilon_C = \delta + \frac{m}{m+n} \, (\epsilon_A - \beta_B) + \frac{n}{m+n} \, \epsilon_B$$

Fig. 4.12   *(m +n)/1 ratio calibration circuit*

4.1.3.4 The (m + n) circuit

The $(m + n)$ circuit shown in Fig. 4.12 provides a means for determining the error of a current comparator whose ratio is equal to the sum of the ratios of two other comparators. Balance is achieved by adjustment of the two primary current sources, the three compensation winding lead voltage sources, the primary Wagner earth circuit, and the measuring network. As with the other circuits, grounding points M_S and M_S' facilitate the adjustment of the compensation winding lead voltage sources.

The balance equation for the $(m + n)$ circuit is derived from the following relationships:

$$I_{SA} = \frac{I_{PA}}{m} \, (1 + \varepsilon_A)$$

$$I_{SB} = \frac{I_{PB}}{n} \, (1 + \varepsilon_B)$$

$$I_{SC} = \frac{I_{PC}}{m + n} \, (1 + \varepsilon_C)$$

$$I_{SC} = I_{SB}(1 + \delta)$$

$$I_{SB}' = I_{SB}(1 + \beta_B)$$

$$I_{SA} = I_{SB}'$$

$$I_{PC} = I_{PA} + I_{PB}$$

from which

$$\varepsilon_C = \delta + \frac{m}{m + n} \, (\varepsilon_A - \beta_B) + \frac{n}{m + n} \, \varepsilon_B$$

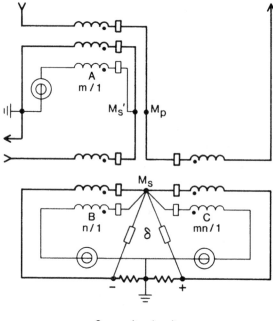

Cascade circuit

$$\epsilon_c = \delta + \epsilon_A + \epsilon_B$$

Fig. 4.13   *(mn)/1 ratio cascade calibration circuit*

4.1.3.5 The mn (cascade) circuit

The *mn* or cascade circuit shown in Fig. 4.13 provides the fastest means for building up to higher ratios. Here the ratio of the comparator being calibrated is equal to the product of the ratio of the two reference comparators. Balance requires the adjustment of the primary Wagner earth circuit, two current sources, three compensation winding lead voltage sources, and the measuring network. As before, grounding points M_S and M_S' facilitate the adjustment of the compensation winding lead voltage sources.

The balance equation is derived from the following relationships:

$$I_{SA} = \frac{I_{PA}}{m}(1 + \varepsilon_A)$$

$$I_{SB} = \frac{I_{PB}}{n}(1 + \varepsilon_B)$$

$$I_{SC} = \frac{I_{PC}}{mn}(1 + \varepsilon_C)$$

$$I_{SC} = I_{SB}(1 + \delta)$$

$$I_{PA} = I_{PC}$$

$$I_{PB} = I_{SA}$$

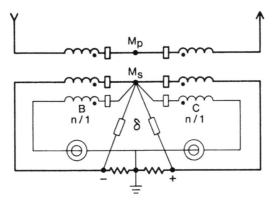

Comparison circuit

$$\epsilon_c = \delta + \epsilon_B$$

Fig. 4.14 *Comparison circuit*

from which

$$\varepsilon_C = \delta + \varepsilon_A + \varepsilon_B$$

4.1.3.6 The comparison circuit
The comparison circuit of Fig. 4.14 enables one comparator to be calibrated against another of the same ratio. It is useful for measuring the effect of the measuring circuit burden on either comparator and for comparing the results of calibrating each comparator individually by the step-up process.

Balance is achieved with four adjustments: the primary Wagner earth circuit controlling the potential of the point M_P, the two compensation winding lead voltage sources (with point M_S grounded), and the measuring network.

4.2 Current transformer calibration circuits

A circuit which uses the compensated current comparator for calibrating a current transformer is shown in Fig. 4.15. It features the supply of energy for the calibration at the secondary current level and automatic maintenance of the measuring points M_P and M_S at ground potential via compensation windings. A ratio error set injects a measured proportion I_c of the secondary current into point M_S from point N. When adjusted to null the detector D, the ratio error set indicates the error of the current transformer I_c/I_S relative to that of the current comparator. $(I_S + I_m)$ is the secondary winding current of the current comparator and $(I_S + I_c)$ is the secondary winding current of the current transformer. I_m is the compensation winding current of the current comparator. It is equal to the magnetising current required by the magnetic shield to establish energy transfer from the secondary circuit to the primary circuit. Since the error

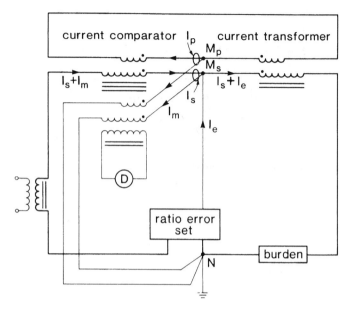

Fig. 4.15   *Current transformer calibration circuit*

of the current comparator is usually much less than that of the current transformer being calibrated, the indication of the ratio error set may be taken
directly as the error of the current transformer.

For ratios involving primary windings with low numbers of turns, the
method of controlling the potential of the primary circuit is relatively inconse-
quential and a direct connection to ground is usually made at some convenient
point in the circuit.

Where a suitable high-current supply transformer is available, energy for the
calibration can be supplied directly to the primary circuit. This reduces the
internal winding potentials of the current comparator and as a consequence the
capacitive components of error. Owing to the uncertainty in this capacitance
effect, operation in this mode is essential for calibrating current transformers
where errors are less than ten parts in 1 million cr so.

In the strictest sense, the currents being compared are defined only at the
points where they enter the nodes M_P and M_S when these nodes are at ground
potential. For the most precise measurements, specific lead pairs (standard
leads) are required since their impedances fix the winding potentials and hence
the capacitance error. The leads should therefore be considered to be an integral
part of a winding.

The burden imposed on the secondary winding of the current transformer
can be readily measured by connecting a suitable resistor across the winding
terminals and noting the change in the ratio error set indication required to
rebalance. Thus for a resistor r and a shift in error of $\Delta\varepsilon$, the burden

$Z = -r \, \Delta\varepsilon$. The resistor r must be large enough so that its application does not appreciably affect the burden Z, yet not so large as to prevent a measurable change in error. In a typical measurement, with $r = 1000$ ohms and the error given in parts per million, the burden is provided directly in milliohms, and the influence of connecting r across a burden Z up to 10 ohms is less than 1%.

4.2.1 The cascade circuit

Attempts to utilise very high ratios in a current comparator merely by increasing the number of turns, either to accommodate large primary currents or milliampere-level secondary currents, will incur an increase in capacitance error.

This may be overcome by connecting two or more comparators in cascade.[3] The current comparator, with its ratio and compensation windings and magnetic shield, is easily adapted to this mode without a significant reduction in its accuracy.

The basic arrangement is shown in Fig. 4.16. It is to be noted that the interconnection between the two comparators is made by connecting the ratio

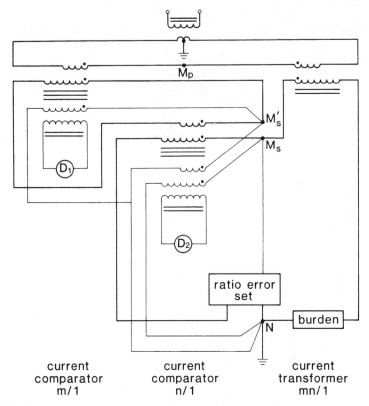

Fig. 4.16 *Cascading current comparators for higher ratio*

Fig. 4.17 *Steps in construction of 60 000 ampere current comparator*

winding outside the magnetic shield in the higher-current comparator to a ratio
winding outside the shield in the lower-current comparator and by making
corresponding connections between the compensation windings inside the
shields of the two comparators. The potentials to ground of the two circuits are
made by the interconnection at point M'_S and the connection of the associated
compensation winding circuit to point N.

For large primary currents where the primary windings involve only one turn
or so, the primary circuit may be grounded at any convenient point. For
cascade circuits used at lower currents, for example to reach milliampere secondary current levels, the point M_P should be brought to ground potential via
a Wagner earth or by including a primary compensation winding in the higher-
current comparator.

In this cascade circuit, the intermediate or higher-current comparator operates as an ideal two-stage current transformer, introducing an error which, for

Fig. 4.18 *A 60 000 ampere current comparator with transformer under test above and power supply transformer below*

most practical purposes, is negligible. This can be readily verified by monitoring the detector D1. However, if a null of detector D1 is required for the utmost accuracy, a small adjustable voltage source can be included in the compensation winding circuit of the higher-current comparator, as shown in Fig. 4.13. Measurement of the current transformer's error is accomplished using the ratio error set to balance detector D2.

A current comparator designed specifically for use at the high-current end of a cascade is shown in various stages of its construction in Fig. 4.17 and completed in Fig. 4.18. The single-turn primary, consisting of horizontal copper plates and ten vertical copper bars and rated at 60 000 amperes, encloses a supply transformer, the current comparator, and the current transformer to be calibrated. The secondary winding, segments of which are distributed in ten coils displaced equally around the toroid, can be connected in series/parallel to realise windings from 120 turns to 1200 turns. Each segment is rated at 50 amperes. The power transformer is driven in series resonance from the supply and requires about 25 kilowatts to generate 60 000 amperes.

4.2.2 Less-than-unity ratios

Current transformers of substantially less-than-unity ratio prove a particular problem because the secondary circuit impedance is reflected into the primary circuit at the square of the turns ratio, and this combines with the winding capacitances to increase the current comparator's error. The solution is to drive the ratio error set from the primary circuit and make the secondary circuit impedance as small as possible.[4] (Conductors whose current rating is far in excess of that actually required should not be dismissed.)

The circuit is shown in Fig. 4.19. Note that the ratio error set must be isolated by a grounded shield from the primary circuit potential. Even though its indication is proportional to the primary circuit, the ratio error set still indicates the error directly because it is associated with primary winding turns.

For most practical purposes, current comparators designed for calibrations at greater-than-unity ratios can also be used at less than unity by simply exchanging the role of their primary and secondary windings. A problem may occur however at very small ratios, where the necessary compensation winding turns are not available. The whole secondary circuit then becomes the burden

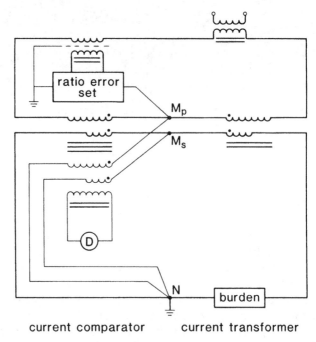

current comparator   current transformer

Fig. 4.19 *Circuit for calibrating current transformer at less than unity ratio*

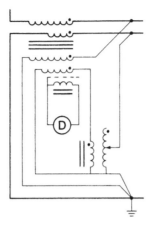

Fig. 4.20 *Adjusting the effective turns of a compensation winding*

on the current transformer and this must be taken into account. Alternatively an auxiliary multiratio current transformer might be used to realise the required effective number of turns, as shown in Fig. 4.20.

4.3 Ratio error sets

Calibration of a current transformer by comparison with a current ratio standard such as a current comparator requires a means to measure the difference between the reference current of the standard and the secondary current induced by the current transformer.[5,6] Since the current ratios of both the standard and the current transformer being calibrated are nominally the same and the secondary circuits are connected in series aiding, the measurement of the difference does not require high precision. Usually an accuracy of 1% or so will suffice.

Ratio error sets for use with current comparators must be of the current injection type; that is, they must have the characteristics of a current source. Typically the voltage at the terminals of a current comparator (M_S and N) may be as high as 5 millivolts. Thus for the ratio error set to have an effect not greater than one part in 1 million, its equivalent source impedance should be not less than 1000 ohms.

A second requirement of a ratio error set is that its input impedance be low so as to minimise the burden imposed on the current comparator.

4.3.1 Admittance-type ratio error sets
The simplest form of ratio error set is the admittance type illustrated in Fig. 4.21. Two variations are shown: that of Fig. 4.21a imposes its burden more or less equally on the current comparator and the transformer being calibrated,

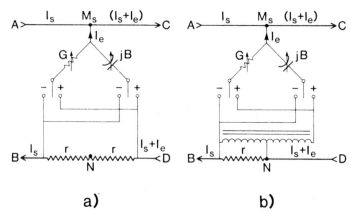

Fig. 4.21   *Resistance/capacitance-type ratio error measuring networks*

and that of Fig. 4.21b employs an autotransformer to transfer the total burden to the current comparator. The current comparator or standard is connected to points A and B while the current transformer being calibrated is connected to points C and D. Points M_S and N are maintained at the same potential with the secondary compensation winding (when comparing two current transformers, achievement of zero potential between points M_S and N indicates balance). At balance, the ratio I_e/I_S or the error ε is, to a first approximation, given by

$$\varepsilon = I_e/I_S \simeq (G + jB)r$$

Approximate values of G, B and r can be chosen to make the set direct reading. The burden imposed by the circuit is essentially equal to the resistance r, usually 0·1 ohm. The equivalent output shunt impedance is given by the parallel combination of the conductance G and the capacitive admittance B.

The principal defect of the admittance-type ratio set is that the current I_e injected into point M_S does not pass through the resistor r and its effect increases with increasing I_e. The correct relationship between the output current I_e and the reference current I_S can be determined by analysing the equivalent circuits shown in Fig. 4.22. Consider Fig. 4.22a:

$$i = -i_1 + i_2$$
$$= -Y_1 r(I - i_1) + Y_2 r(I - i_1)$$
$$= -(Y_1 - Y_2)r(I - i_1)$$

Successively substituting $Y_1 r(I - i_1)$ for i_1 and summing the resulting series,

$$\frac{i}{I} = \frac{-(Y_1 - Y_2)r}{1 + Y_1 r}$$

A similar analysis of Fig. 4.22b yields

$$\frac{i}{I} = \frac{-(Y_1 - Y_2)r}{1 + (Y_1 + Y_2)r}$$

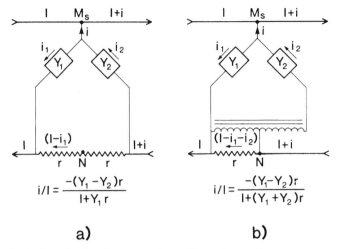

Fig. 4.22 *Equivalent circuits of resistance/capacitance-type ratio error measuring networks*

These two equations illustrate the systematic error in the admittance-type ratio error set. Only if Y_1r in Fig. 4.22a and $(Y_1r + Y_2r)$ in Fig. 4.22b are much less then unity is the desired relationship approached.

This systematic error can be eliminated by separating the in-phase and quadrature components of the circuit and adopting a current-splitting technique. While this is easily accomplished for in-phase component, the quadrature component presents a more difficult problem and a less than satisfactory solution.

4.3.2 Transformer-type ratio error sets

The transformer-type ratio error set, in contrast to the admittance-type set, makes use of fixed admittances and variable transformer ratios. Such a set is shown in Fig. 4.23. It consists of two input current transformers, two three-decade ratio transformers, a phase-shifting network, and an output transformer. The ratios of both the input and output current transformers can be made adjustable for range multiplication.

The principal parameter affecting the design of the ratio error set shown in Fig. 4.23 is the input impedance of the phase-shifting network. When used in the current-shifting mode with a zero output impedance, this input impedance is

$$\frac{2R}{1 + j\omega CR}$$

or

$$(\sqrt{2})R \angle -45$$

at the operating frequency. The associated ratio and current transformers must be designed to support this burden.

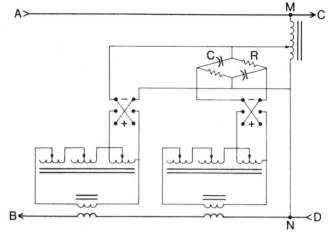

Fig. 4.23 *Transformer-type ratio error measuring network*

The overall input impedance of the set, assuming the various transformer leakage impedances to be negligible, is the input impedance $(\sqrt{2})R$ of the phase-shifting network, modified by the square of the turns ratio of the associated transformers. The output shunt impedance is essentially that of the open-circuit impedance of the phase-shifting network $R/\sqrt{2}$, modified by the square of the turns ratio of the output transformer.

4.3.3 Electronic-type ratio error sets

The functions of a ratio error set can of course also be provided by electronic means. Circuits are readily devised for converting the secondary current into a voltage, introducing in-phase and quadrature scaling, and reconverting to a current output. Of particular interest is a set described by R. L. Kahler[7] which combines a digital panel meter that transforms a DC voltage into a digital output to control a multiplying digital-to-analogue converter and hence the magnitude of the in-phase and quadrature components of the output.

4.4 The phantom burden

The physical simulation of burdens required for the calibration of current transformers where the actual burden is not available is not always easily realised. Resistive burdens, although easy to reproduce, must sometimes be capable of heavy power dissipation, particularly if measurements are to be made at greater than rated current. Inductive burdens of specified power factor, such as those required by national standards, are difficult to design and fabricate.

The burden on a current transformer is defined by the voltage across its

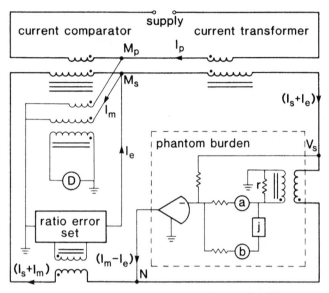

Fig. 4.24 *Current transformer calibration circuit with phantom burden*

secondary winding terminals divided by the secondary current. These parameters relate to the energy transferred from the primary circuit through the current transformer to the secondary circuit where it is usually dissipated in passive components. A more attractive alternative is to have this energy returned to the primary circuit, and the compensated current comparator provides a convenient means for achieving this.[8]

A current transformer calibration circuit with the phantom burden installed is shown in Fig. 4.24. It is to be noted that, although the compensation windings remain connected to ground, the point N is not. By driving the point N to a particular voltage with an amplifier it is possible to impose a desired burden on the current transformer and an equal but negative burden on the current comparator. Thus energy can be transferred through the current transformer from the primary to the secondary circuit and returned back to the primary through the secondary winding of the current comparator. In the circuit as shown, a voltage developed by the current $(I_S - I_e)$ and the resistor r is modified by controls a and jb and then compared with the voltage V_S, yielding

$$V_S = K(a + jb)(I_S - I_e)r$$

or a burden

$$Z = V_S/(I_S - I_e) = K(a + jb)r$$

The burden on the current transformer is thus defined by the voltage between point M_S, which is at ground potential, and the point V_S, which is at V_S potential, as set by controls a and jb. In order to control this circulation of energy the amplifier is required to deliver the magnetising currents of the current transformer and the current comparator at the desired burden voltage. Further, the output impedance must be sufficiently low to meet the desired overall accuracy. Since the magnetisation characteristics are non-linear, these two factors become critical at different points in the overall operating range.

The volt-ampere requirement of the amplifier is given by

$$S = V_S(I_m - I_c)$$

$$\simeq I_S^2 Z \left(\frac{I_m}{I_S} - \frac{I_c}{I_S} \right)$$

$$\simeq I_S^2 Z \left(\frac{Z}{Z_m} - \varepsilon_{ct} \right)$$

where

$I_S + I_c$ = secondary current of the current transformer being calibrated, I_S being the nominal or reference secondary current and I_c being the error current

Z = burden impedance

Z_m = magnetising impedance of the secondary winding of the current comparator with respect to the magnetic shield

ε_{ct} = error (I_c/I_S) of the current transformer being calibrated

In general an accuracy of 1% is sufficient for generating a burden; hence a feedback loop gain of 100 is satisfactory.

4.5 Calibration of two-stage current transformers

Two-stage current transformers are constructed with two magnetic cores, both of which are embraced and excited by the primary and secondary windings. A third (auxiliary) winding, normally having the same number of turns as the secondary winding, embraces and acts with only one of the two cores. A compensated current comparator without the detection winding is essentially a two-stage current transformer. However many two-stage current transformers employ a side-by-side core configuration rather than the concentric toroid approach of the current comparator. This leads to slightly less but nevertheless still very high accuracy and the two-stage current transformer is very useful in extending the range of precision instrumentation.

Two-stage current transformers are either used with individual burdens, such as in the cascade circuit of Fig. 4.16, or with a small common burden. Their errors are sensitive to the voltage of the second stage and strict control of the conditions under which the measurements are made is imperative. In particular the potential of the point M_S must be the same as point N and equal to ground. This can be achieved by using an auxiliary voltage source in the compensation winding lead, as shown in Fig. 4.7 and following the two-step balancing procedure outlined in Section 4.1.3.1.

References

1 N. L. Kusters and W. J. M. Moore. 'The compensated current comparator; a new reference standard for current transformer calibrations in industry.' *IEEE Trans. Instrumentation and Measurement*, **IM-13**(2–3), June–September 1964, pp. 107–14

2 N. L. Kusters and W. J. M. Moore. 'The development and performance of current comparators for audio frequencies.' *IEEE Trans. Instrumentation and Measurement*, **IM-14**(4), December 1965, pp. 178–90

3 P. N. Miljanic, N. L. Kusters, and W. J. M. Moore. 'The application of current comparators to the calibration of current transformers at ratios up to 36 000/5 amperes.' *IEEE Trans. Instrumentation and Measurement*, **IM-17**(3), September 1968, pp. 196–203

4 N. L. Kusters and W. J. M. Moore. 'The application of the compensated current comparator to the calibration of current transformers at ratios less than unity.' *IEEE Trans. Instrumentation and Measurement*, **IM-18**(4), December 1969, pp. 261–5

5 P. N. Miljanic. 'Adjustable complex-ratio transformer.' *IEEE Trans. Instrumentation and Measurement*, **IM-14**, September 1965, pp. 135–41

6 W. J. M. Moore and N. L. Kusters. 'Direct reading ratio-error sets for the calibration of current transformers.' *IEEE Trans. Instrumentation and Measurement*, **IM-19**(3), August 1970, pp. 161–6

7 R. L. Kahler. 'An electronic ratio error set for current transformer calibrations.' *IEEE Trans. Instrumentation and Measurement*, **IM-28**(2), June 1979, pp. 162–4

8 N. L. Kusters and W. J. M. Moore. 'A phantom burden for current transformer calibration.' *IEEE Trans. Power Apparatus and Systems*, **PAS-93**(1), January/February 1974, pp. 240–3

Chapter 5

High-voltage capacitance bridges

Transformer ratio bridges may be configured in two ways. In the voltage bridge, the transformer is used to determine the ratio of the voltages across the other two arms when the same current passes through them. In the current bridge the transformer is used to determine the ratio of the currents in the other two arms when the same voltage is applied to them. The voltage bridge operates with a finite flux in the magnetic core while in the current bridge the flux in the core is ideally at zero. The transformers in precision bridges are usually of the three-winding type, the extra winding being used to excite the core in the voltage bridge or to detect the zero flux, and hence the ampere-turn balance condition, in the current bridge.

Fundamentally there is no basic difference between the two types of bridge. One can be obtained from the other merely by exchanging source for detector. However in the voltage bridge there are limits on the flux magnitude owing to core saturation and this places some constraints on core size, the numbers of turns, and the operating frequency. For use with high-voltage capacitors, the current bridge is essential.

The three-terminal gas-dielectric capacitor is unique in providing a relatively stable and loss-free reference for use at high voltages. For many years its full potential could not be realised because of difficulties with the low-voltage ratio arms. Resistors in this application suffered from heating and phase angle errors, and usually a second balance was necessary to screen the detector. Transformer ratio arms have eliminated many of these problems, particularly the need for the screen balance, and the current comparator with its magnetic shielding is ideally suited to this application.

5.1 Configuration and design considerations

5.1.1 Basic bridge
The basic configuration of a capacitance bridge[1,2] for high-voltage applications is shown in Fig. 5.1. It has two main windings, N_X and N_S, both usually having

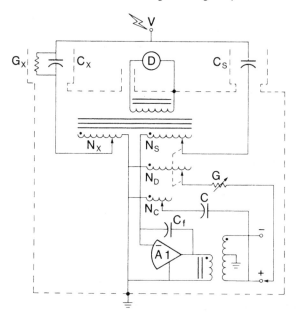

Fig. 5.1 *High-voltage current-comparator-based capacitance bridge*

about 1000 turns. The N_x winding, which is connected to the capacitor C_x to be measured, is used for range multiplication, having taps at say 1, 2, 5, 10, 20, ..., 500, and 1000 turns. The N_s winding, which is connected to the reference capacitor, is wound so that single-turn increments are available. The capacitors C_s and C_f together with the operational amplifier A1 form a capacitive divider, the result being that the output voltage of the amplifier is a negative replica of the applied voltage V, reduced in magnitude by the ratio C_s/C_f. This reduced voltage is used with conductance G to provide phase balance (ganging turns N_D with turns N_s makes the indication direct reading in dissipation factor or 'tangent δ'), and also with capacitance C to further subdivide the main balance into an additional 1000 parts, yielding an overall resolution of one part in 1 million.

To avoid interference from ambient electric and magnetic fields, the bridge is completely enclosed in a metal container and the current comparator itself in a shield can of Mumetal or other magnetic alloy of adequate thickness. Rotary switches, with metallic shafts capable of being grounded, should be used. External coaxial cables should be of the double-shield variety.

The sensitivity of the bridge is determined by the ampere-turns and the magnetising impedance of the ratio windings acting on the inner core. The lower bridge arm potentials are however at balance determined by the winding leakage impedances. Hence it is generally acceptable to connect all screens directly to ground, thus eliminating the need for a screen balance. (The influence of additional screen capacitance, such as may occur when connecting to a capacitor at some distance from the bridge through a coaxial cable, may be tested by connecting a known capacitance from the end of the winding to ground and observing the change in the balance setting.)

The balance equation of the bridge is

$$V(G_X + j\omega C_X)N_X = Vj\omega C_S N_S + V\frac{C_S}{C_f}(j\omega C N_C + G N_D)$$

where

ω = angular frequency in radians per second
C_X = capacitance being measured
G_X = equivalent parallel loss conductance of the capacitance
C_S = reference capacitance, assumed to have no loss

and from which, with $N_D = N_S$,

$$C_X\left(1 + \frac{G}{j\omega}\right) = C_S\frac{N_S}{N_X}\left(1 + \frac{C}{C_f}\frac{N_C}{N_S} + \frac{G}{j\omega C_f}\right)$$

Ignoring the influence of the additional capacitance balance CN_C/C_fN_S, which has a maximum value of 0·001 11, the quantity $G/j\omega C_f$ becomes equal to the dissipation factor or 'tangent δ' of the unknown capacitance C_X (where δ is the loss angle).

The value of capacitor C_f is chosen taking into account the current rating of the reference capacitor and the volt-ampere output capability of amplifier A1. A high-quality polystyrene dielectric capacitor of 1 to 3 microfarads is typically selected since at most 0·1% accuracy overall is acceptable. Although the C/N_C capacitor/winding combination is required to have three-digit resolution, it is not necessary (or desirable if internal winding capacitance error is to be held to a minimum) for the winding N_C to have 1000 turns. A single 100-turn winding, tapped at every 10 turns, can for example be used with multiple switching and a set of capacitors varying in magnitude by factors of ten. A similar arrangement can be made for the dissipation factor balance. To accommodate a measurement capability of 10%, however, five-digit resolution must be available as well as a provision for ganging the three most significant digits with those of the main N_S winding.

With 1000-turn windings the maximum nominal ratio between the measured capacitance and the reference is 1000/1 which inhibits the measurement of large industrial capacitors using gas-dielectric references. This ratio can be extended however by cascading a two-stage current transformer into the N_X winding as shown in Fig. 5.2. Since this winding presents a common burden to both the secondary and auxiliary windings of the current transformer, two sets of interconnecting leads are used and the N_X winding is set to a minimum (typically 20 turns or less) to avoid imposing a large burden on the current transformer and thus compromising the overall accuracy.

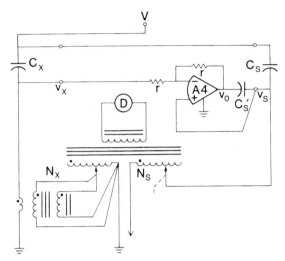

Fig. 5.2 *Ratio extension and circuit for lead and winding impedance compensation*

The principal factor affecting the overall accuracy of the bridge is the capacitance of the 1000-turn windings. Some adjustment can be effected by connecting a suitable capacitance in shunt with, or to ground from, one of the windings. This correction is limited however by the variations imposed by each ratio.

A systematic error is present in the bridge which is significant at large tangent δ settings ($\geqslant 0\cdot001$). Phase imperfections in the amplifier A1 which drives the tangent δ circuitry can affect the three least significant digits of magnitude (N_S) settings.

5.1.2 Improved bridge
A more direct method of reducing the capacitance error and achieving higher accuracy is to use windings of reduced numbers of turns with range-extending two-stage current transformers, and to apply the compensation winding technique to reduce the potential across the winding with the highest number of turns. An improved bridge[3] using these techniques to obtain an accuracy increase of an order of magnitude or so is shown in Fig. 5.3.

In this bridge the main ratio windings have 100 turns. The ratio capability is increased from 100/1 to 10000/1 by the transformer T1 and the resolution of the N_S winding is augmented to one part in 10 million by a five-decade inductive divider T2. Although not shown as such, both the transformer T1 and the two most significant decades of the inductive divider T2 are of two-stage design. Since winding N_S is usually operated close to its maximum number of turns, it is connected in parallel with a compensation winding of the same maximum number of turns located inside the magnetic shield. This reduces the turn-to-turn potentials in the N_S winding and hence its shunt capacitance error.

To avoid adding an additional winding and its capacitances, the dissipation

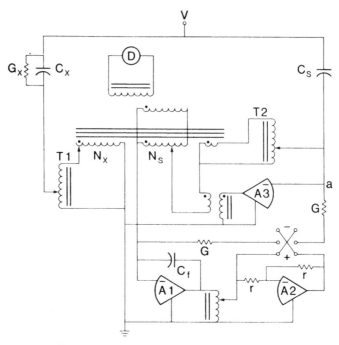

Fig. 5.3 *Improved high-voltage capacitance bridge*

factor balance current is superimposed on the capacitive reference current in the N_S winding. This is accomplished with capacitor C_f acting with the operational amplifier A1 to form a capacitive voltage divider which produces a replica of the applied voltage V reduced in magnitude by the ratio C_S/C_f. This voltage is then used with inverting amplifier A2 and the two conductances G to generate and reclaim the dissipation factor balance current. Amplifier A3 and its associated output transformer act to bring point a to ground potential, thus eliminating the effect of the impedances of the inductive divider T2 and the N_S winding. As in the basic bridge, the ratio can be further extended to $10^7/1$ using an additional two-stage current transformer.

5.1.3 Winding voltage correction circuit
A further refinement of the capacitance bridge, which is essential where low capacitive impedances are involved, is the correction circuit which reduces the effect of the lower-arm leakage impedances on the measurement accuracy. One such circuit is shown in Fig. 5.2. Here the circuit with amplifier A4 acts to inject a correcting current into the N_S winding to overcome the lead impedance voltages v_X of the C_X circuit and v_S of the C_S. Then the bridge balance equation becomes

$$(V - v_X)j\omega C_X N_X = (V - v_S)j\omega C_S N_S + (v_0 - v_S)j\omega C'_S N_S$$

and since $v_0 = 2v_s - v_x$ and $C_X N_X \simeq C_S N_S$, and if $C'_S = C_S$,

$$V j\omega C_X N_X = V j\omega C_S N_S$$

To complete the four-terminal connection to the C_X capacitor, the high-impedance reference capacitor is connected directly to its high-voltage terminal.

Although it is not always necessary to correct for the lead impedances of the C_X capacitor, the correction circuitry for the N_S winding is usually kept as a permanent fixture.

5.2 Capacitance bridge calibration

5.2.1 Capacitance ratio dials

Capacitance bridges can be calibrated using a scale of gas-dielectric capacitors with substitution and build-up techniques.[4] For this purpose it is convenient to have capacitors that can be trimmed to exact values to avoid the necessity of having to manipulate minute corrections.

The procedure begins at the 1/1 ratio setting of the N_X winding where three nominally equal capacitors of say 50 pF are trimmed to exact equality by substitution, and the bridge balance point is established by interchanging the capacitors between the C_X and C_S positions. The 2/1 ratio is then calibrated by connecting two 50 pF capacitors in the C_X position, and the trim of the 100 pF capacitor established by substitution. Similar steps are followed with additional capacitors for the higher ratios. At the higher end of the ratio scale, from say 100/1 to 1000/1, the available capacitance ratio becomes somewhat limited by sensitivity and size considerations, and it is sometimes useful to employ a 10/1 ratio three-winding voltage transformer to augment the capacitance ratio. Steps must be taken of course to eliminate this voltage transformer's error from the measurements. And, since the actual capacitance of the capacitors may vary with the applied voltage, the same applied voltage should be used throughout the build-up procedure.

The magnitude (N_S) dial settings can be calibrated in a similar manner. It is not of course necessary to maintain the same resolution and accuracy for each dial. However it is necessary in the basic bridge to keep the most significant digit at a high setting while calibrating the least three significant digits, since the dissipation factor balance is not effective without a reasonable number of main winding turns in circuit.

Where feasible, calibrations should be made with ratios of large capacitances as well as smaller ones to ascertain the effect of loading. Alternatively, this effect can be determined by connecting capacitances to ground from the same points where the C_X and C_S capacitors are connected, and noting the change in balance.

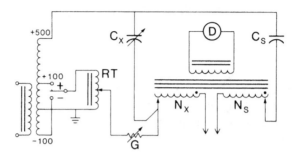

Fig. 5.4 *Circuit for calibrating the dissipation factor scale*

5.2.2 Dissipation factor dials

The accuracy of the dissipation factor measurement is limited by the feedback amplifier A1 to about 0·5% of reading; hence the calibration circuitry is less demanding than that required for the capacitance ratio dials. A suitable arrangement is shown in Fig. 5.4.

In this circuit C_S is a standard reference capacitor of 1000 pF, the bridge is set to exactly unity ratio, and the capacitor C_X is adjusted for magnitude (capacitance) balance. Measured amounts of quadrature current are then injected into the N_X winding through conductance G. G is a conductance decade ranging from 0·1 to 1·0 microsiemens. Appropriate settings of the ratio transformer RT are chosen for each dissipation factor decade switch being calibrated.

Actual measurement of the dissipation factor dial errors can usually be made with a decade of lesser significance. If the size of the error is incompatible with the measuring dial, an offset can be made in the setting of the ratio transformer RT.

5.3 Capacitance and dissipation factor measurement

The accurate measurement of capacitance and dissipation factor using bridge techniques requires close attention to interconnections and layout. All coaxial leads must be of the double-shield variety. Care should be taken also to avoid unnecessarily high currents in these cable shields since potential differences between the shield and inner conductor can introduce unwanted dissipation factor shifts. Additional large-size conductors should be used to connect third-terminal capacitances to ground. Where two-stage current transformers are involved, two individual pairs of leads, enclosed in a shield, should be used to minimise the common burden.

When the bridge is operated in an industrial environment it is good practice to test for the unwanted influence of ambient electric and magnetic fields. To accomplish this the high-voltage connections are removed and the high-voltage terminals of the capacitors are connected to ground. The voltage being applied to the test object is then raised to the measuring level and the deflection on the bridge detector noted. This deflection sets the limit of resolution in the measurement.

Alternatively, if the C_X measurement involves large currents and a two-stage range-extending transformer, the ambient field test should be made by arranging for the C_X current to bypass the primary winding of the range extender. In this way the influence of the current is not omitted.

The measured capacitance is obtained directly from the settings of the N_S and N_X windings, the ratio of any range-extending two-stage current transformer, and the value of the reference capacitor. It is frequency insensitive. The dissipation factor, obtained from a calibrated conductance, is however frequency specific. The dissipation factor dials, scaled for a frequency of f_0, will indicate what the dissipation factor would be at frequency f_0 even though the measurement is made at another frequency f. If the dissipation factor is required for the frequency f, the dial indication must be multiplied by the factor f_0/f.

5.4 Reactor and transformer loss measurement

The capacitance bridge with a two-stage current transformer can be used to measure the inductance and loss angle of shunt reactors and power transformers on short-circuit.[5,6]. The features of the bridge that make this possible are the ability to reverse the phase of the reactor/transformer current and to measure what is essentially a negative phase angle. With an impedance $(R_X + j\omega L_X)$ replacing the capacitor C_X, and the polarity of the primary winding of the current transformer reversed (see Fig. 5.5), the capacitance bridge equations become

$$L_X = 1/mn\omega^2 C_S$$

$$R_X = G/mn\omega^2 C_S C_f$$

$$\tan\delta = R_X/\omega L = G/\omega C_f$$

where

m = ratio of the range-extending current transformer
n = bridge ratio N_S/N_X
C_S = reference capacitance
C_f = feedback capacitance
ω = angular frequency in radians per second

The power loss P is given by

$$P = V^2 mn\omega C_S \tan\delta \approx I^2 \tan\delta/mn\omega C_S$$

It is important to note here that the bridge balance parameters are in themselves

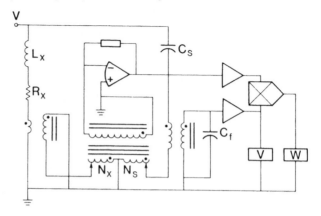

Fig. 5.5 *Semi-automatic balancing reactor/transformer loss bridge*

not sensitive to small voltage fluctuations, so the values for the voltage V and the current I can be those for which the loss measurement was requested and not the actual values encountered in the measurement.

The tan δ balance of the bridge is independent of frequency but the dial indication is specific to that for which it was calibrated and not the frequency of measurement. The main inductive balance however is sensitive to the square of the frequency. This imposes a requirement for short periods of relatively stable frequency and a sinusoidal test voltage waveform of low distortion to enable a balance to be obtained. Residual harmonic voltages at the detector can result in small differences in the balance point depending on whether tuned or gated detectors are used.

A solution to the frequency problem is to determine the deviation of the test frequency from nominal and then inject a proportional correction current into the reference (C_S) side of the bridge.[7] Alternatively, semi-automatic balancing can be provided,[8] as shown in Fig. 5.5. However the output indication of this bridge, which is directly proportional to power loss, is sensitive to the square of the voltage fluctuation, so that any loss measurement must be accompanied by a simultaneous voltage measurement.

The measurement of loss in three-phase transformers poses additional problems because of the inaccessibility of phase currents at the neutral or low-voltage end of the windings. Special input transformers, insulated to withstand the short-circuit impedance voltage at the high-voltage end of the windings must be provided. Only one ground should be connected to the system, preferably at the neutral point of the transformer, to avoid neutral currents. The neutral point of any power factor correcting capacitors should be isolated. It is difficult if not impossible to realise physical coincidence between the bridge ground and the electrical neutral point of the transformer; hence individual phase loss measurements have little practical meaning and only the total of all three phase loss measurements can be relied upon.

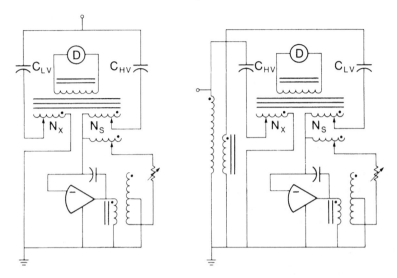

Fig. 5.6   *Voltage transformer calibration*

5.5 Voltage transformer calibration

The capacitance bridge can be used for the calibration of voltage transformers in a two-step process as illustrated in Fig. 5.6. The bridge is first used to establish the ratio between low-voltage and high-voltage capacitors C_{LV} and C_{HV}. To avoid the need to carry forward very precise corrections it is convenient to have the low-voltage capacitor C_{LV} adjustable so that it can be trimmed to an exact integral ratio. The second step of the process is to exchange the connections of the two capacitors to the bridge windings and apply the two voltages whose ratio is to be determined to the appropriate capacitors. The setting of the N_S winding, normalised to unity, gives the ratio correction factor and the dissipation factor dials yield the phase angle error in radians. (The ratio correction factor is defined as the actual ratio divided by the nominal ratio. The phase angle error is the angle by which the actual secondary voltage leads the nominal secondary voltage.)

A requirement of the measurement method is that the capacitance of the high-voltage capacitor remains constant with variation of the applied voltage.

Voltage transformers are usually calibrated with zero burden and with a known resistive burden. From these two measurements, the errors at any other burden can be calculated using the following equations:[9]

$$RCF_\ell = RCF_0 + \frac{r}{Z}[\Delta RCF \cos\theta - \Delta\phi \sin\theta]$$

$$\phi_\ell = \phi_0 + \frac{r}{Z}[\Delta\phi \cos\theta + \Delta RCF \sin\theta]$$

where

Z, θ = impedance and phase angle of the burden for which the calibration is desired

RCF_0, RCF_z = ratio correction factors at zero burden and at the desired burden Z, respectively

ϕ_0, ϕ_z = phase angle errors in radians at zero burden and at the desired
burden, respectively

r = resistive burden

$\Delta RCF, \Delta\phi$ = values obtained at burden r minus those obtained at zero
burden

If the burden connected to the voltage transformer is not known, it can be
measured with the bridge. To accomplish this a small resistor, whose resistance
is negligible with respect to the burden impedance Z, is connected in series with
the high-voltage terminal of the burden. Bridge balances are then obtained with
the low-voltage capacitor C_{LV} connected to the higher-voltage end of the resistor (position a) and then to the lower-voltage end (position b). The admittance
Y of the burden Z is then given by

$$Y \simeq (M_{b-a} + jD_{b-a})/r$$

where $M_{b\ a}$ is the value of the RCF at position b minus that at position a, and
$D_{b\ a}$ is the value of the phase angle error at position b minus that at position
a. To avoid second-order errors the value of r should be such that the magnitudes of $M_{b\ a}$ and $D_{b\ a}$ are not greater than 0.001.

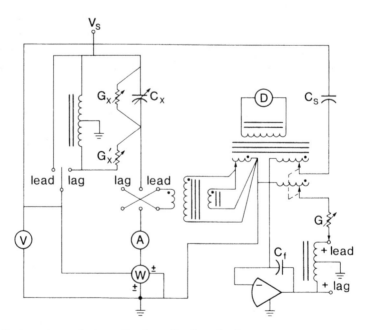

Fig. 5.7 *Low-power-factor wattmeter calibration circuit*

5.6 Low-power-factor wattmeter calibration

The accurate measurement of power at low power factors depends almost entirely on the phase angle errors of the measuring equipment. The capacitance bridge enables the determination of these errors in wattmeters by providing the means for calibrating them under zero-power-factor conditions.[10] The circuit is shown in Fig. 5.7.

To realise a 5 ampere current and 120 volts, adjustable phase-locked power supplies are required. Alternatively a 5 ampere current at 120 volts can be obtained as shown in Fig. 5.7 with a capacitive impedance of 24 ohms or a capacitance of approximately 110 microfarads. Such capacitors have fairly large dissipation factors, represented by the conductance G_X. Compensation is obtained by providing a negative in-phase voltage which together with an adjustable conductance G'_X can be used to extract the loss current from the circuit before it enters the bridge. Inclusion of some air-core inductance in series with the voltage source so that the system can be operated near series resonance ensures good sinusoidal waveform conditions and improves the power factor.

References

1 N. L. Kusters and O. Petersons. 'A transformer-ratio-arm bridge for high-voltage capacitance measurements.' *IEEE Trans. Communications and Electronics*, 69, November 1963, pp. 606–11

2 O. Petersons. 'A transformer-ratio-arm bridge for measuring large capacitors above 100 volts.' *IEEE Trans. Power Apparatus and Systems*, **PAS-87**(5), May 1968, pp. 1354–61

3 O. Petersons and W. E. Anderson. 'A wide-range high-voltage capacitance bridge with one PPM accuracy.' *IEEE Trans. Instrumentation and Measurement*, **IM-24**(4), December 1975, pp. 336–44

4 W. K. Clothier and L. Medina. 'The absolute calibration of voltage transformers.' *J. IEE*, **104**, part A, June 1957, pp. 204–11

5 W. J. M. Moore and F. A. Raftis. 'Measurement of shunt reactor loss at high voltage with an alternating current comparator bridge.' *IEEE Trans. Power Apparatus and Systems*, **PAS-92**(2), March April 1973, pp. 662–6

6 W. J. M. Moore, G. Love, and F. A. Raftis. 'Measurement of short circuit load losses in large three phase power transformers using an alternating current comparator bridge.' *IEEE Trans. Power Apparatus and Systems*, **PAS-94**(6), November December 1975, pp. 2074–6

7 E. So. 'An improved frequency-compensated capacitance bridge for accurate shunt reactor loss measurements at very low power factors.' *IEEE Trans. Power Apparatus and Systems*, **PAS-103**(5), May 1984, pp. 1099–103

8 R. Malewski, W. E. Anderson, and W. J. M. Moore. 'Interlaboratory comparison of EHV shunt reactor loss measurement.' *IEEE Trans. Power Apparatus and Systems*, **PAS-99**(4), July/August 1980, pp. 1634–41

9 J. L. Settles, W. R. Farber, and E. E. Connor. 'The analytical and graphical determination of complete potential transformer characteristics.' *IEEE Trans. Power Apparatus and Systems*, 79(51), February 1961, pp. 1213–18

10 W. J. M. Moore. 'A technique for calibrating power frequency wattmeters at low power factors.' *IEEE Trans. Instrumentation and Measurement*, **IM-23**(4), December 1974, pp. 318–22

Chapter 6

Low-voltage bridges

6.1 Power bridge

A unique standard for the unit of power does not exist. The measurement of power must rely on a reference that is derived from the standards of voltage and resistance. Various techniques have been developed for combining these two standards in such a way as to provide a practical and convenient power standard.[1,2] The power bridge, based on the current comparator, is one way of achieving this objective.

The bridge[3] is shown schematically in Fig. 6.1. The current I entering winding N_X on the current comparator is compared with an in-phase current in winding N_R obtained by applying the voltage V to resistor R and with a quadrature current in winding N_C obtained by applying the voltage V to capacitor C. At ampere-turn balance of the current comparator,

$$IN_X = \frac{V}{R} N_R \pm j\omega C V N_C$$

where N_X, N_R and N_C are the actual number of active turns in their respective windings.

Rearranging and multiplying by the voltage V,

$$VI = \frac{1}{N_X} \left(\frac{V^2}{R} N_R \pm j\omega C V^2 N_C \right)$$

The bridge thus resolves the apparent power VI into an active power component proportional to V^2/R and a reactive power component proportional to $\omega C V^2$, where V and I are the root mean square voltage and current respectively.

In the actual bridge, the equivalent of six-digit resolution is obtained in the N_R and N_C windings by cascading 100-turn single-stage current transformers, which provide the two least significant digits, into 100-turn two-stage current transformers for the middle pair of digits, and thence to 100-turn windings on the current comparator for the two most significant digits (see Fig. 6.2). The N_X winding is configured for ratio multiplication, having taps at 1, 2, 5, 10, 20, 50

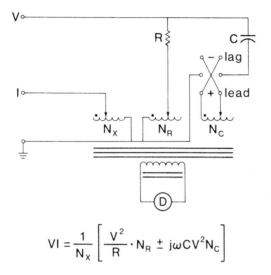

$$VI = \frac{1}{N_X}\left[\frac{V^2}{R}\cdot N_R \pm j\omega CV^2 N_C\right]$$

Fig. 6.1 *Basic power measuring circuit*

and 100 turns. With a current rating of 1 ampere-turn for the current comparator windings, the bridge can be operated at 120 volts, 1 ampere, using resistive and capacitive impedances of 12 000 ohms at 60 Hz. For 5 ampere operation, the windings on the current comparator can be increased to 500 turns, which increases its overall size and magnifies the capacitance error. Alternatively a 5/1 two-stage current transformer can be connected to the N_X winding.

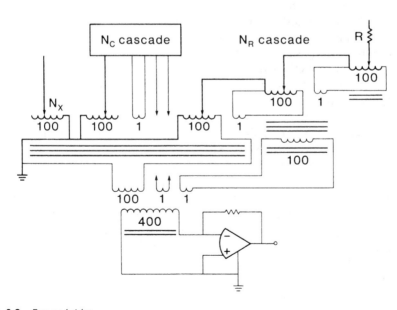

Fig. 6.2 *Power bridge*

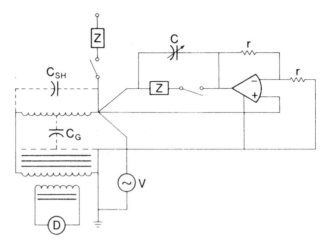

Fig. 6.3 *Adjustment of capacitive error correction*

Since the resistor R and capacitor C are used to define the currents in the N_R and N_C windings, corrections must be made for the impedances of the windings themselves. This is achieved with circuits similar to that shown in Fig. 5.2. For the purposes of this section, the connection is made not to the low-voltage terminal of C_X but to ground, and the capacitor C'_S is replaced by a resistor for the in-phase cascade.

The compensation winding on the current comparator can also be used to assist in reducing the voltage across a ratio winding, but only one of these at a time. It can be switched between either the N_R or the N_C windings depending on which has the largest number of active turns.

With a compensation winding and a winding impedance voltage correction circuit in place, means are provided to compensate for the capacitance error of the winding. The method of adjustment is shown in Fig. 6.3. The winding voltage distribution in the bridge configuration differs from that of the current transformer so that the effect of the shunt capacitance C_{SH} has the same sign as that of the capacitance to ground C_G and compensation must be provided by an external capacitance C. This capacitor is adjusted to null the detector D for reasonable variations in the voltage V.

The power bridge divides the apparent power into two orthogonal components – the active power and the reactive power. Provided that the orthogonality is maintained, inaccuracy in one of the components does not affect the other.

When used in a calibration system[4] the current comparator can be connected in a feedback arrangement to control the magnitude and phase of the current in accordance with the bridge settings N_X, N_R and N_C. This, together with the voltage, establishes the measurement conditions and makes possible the calibration of watt/watt-hour, var/var-hour, volt-ampere-hour, ampere-squared-hour and other similar types of meters.

A limitation of the bridge configuration shown in Figs. 6.1 and 6.2 arises from the use of a capacitor for obtaining a quadrature reference current. Although operation at test frequencies of 50 or 60 Hz can readily be accommodated by adjusting the turns in the N_C winding, extension to frequencies of 400 or 1000 Hz would require a corresponding reduction in the capacitance to avoid excessive currents. A more satisfactory alternative is to use two digital oscillators phase locked at 90 degrees and resistive current references. Such a bridge would then be independent of frequency.

6.2 AC resistance bridge

If the voltage (or some known multiple of the voltage) is applied to a resistor and the resulting current is passed through the N_X winding, the power bridge can be used to measure the magnitude and phase of that resistor.[5] Auxiliary circuits to compensate for the N_X winding impedance are of course required, so that two bridges are not directly interchangeable.

With the resistor to be measured connected to winding N_X, the bridge is direct reading in conductance, with the phase defect measured as an equivalent parallel capacitance. By connecting the unknown resistor to the N_S winding and the reference resistor to the N_X winding, the bridge becomes direct reading in resistance but the phase angle is no longer direct reading in equivalent parallel capacitance. (The capacitive balance acts to modify the phase defect of the reference resistor; hence it is proportional to the negative of the equivalent parallel capacitance of the unknown resistor.)

6.3 Active capacitor/quadrature current reference

The application of the power bridge to the measurement of reactive power requires a stable and accurate quadrature current reference. Gas-dielectric capacitors meet the accuracy and stability requirements but their impedances at power frequencies are too high for practical use. Solid-dielectric capacitors have phase angle errors which are relatively easy to correct[3] but their magnitudes are not sufficiently stable. A solution to this problem is to use a self-balancing current comparator to determine the error.[6]

The basic circuit is shown in Fig. 6.4. Capacitor C_S is a gas-dielectric capacitor of, for example, 1000 pF connected to windings N_2 and N_2', both of which have the same number of turns N_2. Winding N_2' acts as a compensation winding in parallel with winding N_2, maintaining point M at ground potential. Capacitor C_S' is a solid-dielectric capacitor connected to winding N_1 and load Z, having approximately 265 times the capacitance of C_S and hence an impedance at 60 Hz of 10 000 ohms. The ratio of the windings N_2/N_1 is 265. Winding N_D detects any ampere-turn imbalance in windings N_1, N_2 and N_2', and through the

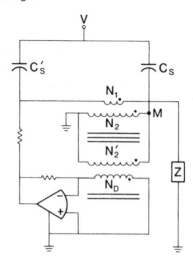

Fig. 6.4 *Basic active capacitor/quadrature current reference*

amplifier injects a correction current into winding N_1 and thence to the load Z. The current in the load Z, which represents a winding on the main current comparator, is thus controlled by the gas-dielectric reference capacitor C_S.

With an electronic integrating circuit replacing the solid-dielectric capacitor C_S', and appropriate changes in the polarities of the N_1 and N_D windings, the circuit can also be used to provide a quadrature current source with the frequency characteristics of an inductance.

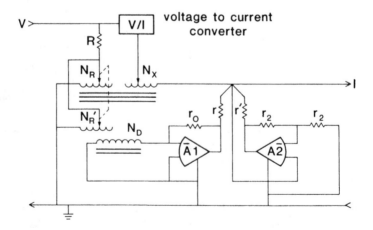

Fig. 6.5 *Transconductance amplifier*

6.4 Transconductance amplifier

A requirement of a transconductance amplifier in precision measurements is that it faithfully reproduces the input voltage waveform as a current. This is usually achieved by measuring the output current with a resistive shunt and comparing the shunt voltage drop with the input. However it is difficult to construct such a shunt for larger currents which has a satisfactory magnitude stability and low phase angle error. The current comparator provides a means to overcome this problem, using its very accurate ratio characteristics to enable a high-valued resistor with very small errors to be used as a reference.

The circuit of the transconductance amplifier is shown in Fig. 6.5. It consists of a reference R connected between the voltage input V and the current comparator ratio winding N_R with its associated compensation winding N'_R, a voltage-to-current converter connecting the voltage V to the current comparator ratio winding N_X, and the current comparator detection winding N_D which senses the ampere-turn imbalance and with the aid of amplifier A1 corrects the output current I through resistor r. Amplifier A2 with its associated circuitry adds additional current to the output to compensate for the possibility that the injection point may not be at ground potential (see Section 5.1.3). The compensation winding N'_R reduces the influence of the N_R winding impedance but, if the reduction is found to be insufficient, a correction circuit similar to that employing amplifier A2 could be used there as well.

The transconductance amplifier thus provides a current I whose waveform is a faithful replica in both amplitude and phase of the voltage V, as determined by the current in the high-quality resistor R. It should be noted that the amplifier is not restricted to sinusoids only but can be used with distorted waveforms as well. By applying direct current comparator techniques, a direct current transconductance amplifier can also be realised.

References

1 W. J. M. Moore. 'A technique for calibrating power frequency wattmeters at low power factors.' *IEEE Trans. Instrumentation and Measurement*, **IM-23**(4). December 1974, pp. 318 22

2 K. J. Lentner. 'A current comparator system to establish the unit of electrical energy at 60 Hz.' *IEEE Trans. Instrumentation and Measurement*, **IM-23**(4), December 1974, pp. 334–6

3 W. J. M. Moore and K. Ayukawa. 'A current comparator bridge for power measurement.' *IEEE Trans. Instrumentation and Measurement*, **IM-25**(4), December 1976, pp. 550 3

4 W. J. M. Moore and E. So. 'A current-comparator-based system for calibrating active reactive power and energy meters.' *IEEE Trans. Instrumentation and Measurement*, **IM-32**(1), March 1983, pp. 147 9

5 E. So and W. J. M. Moore. 'A direct-reading AC comparator bridge for resistance measurement at power frequencies.' *IEEE Trans. Instrumentation and Measurement*, **IM-29**(4), December 1980, pp. 364 6

6 E. So and W. J. M. Moore. 'A stable and accurate current-comparator-based quadrature-current reference for power frequencies.' *IEEE Trans. Instrumentation and Measurement*, **IM-31**(1), March 1982. pp. 43 6

Chapter 7

Direct current comparators

A unique feature of the current comparator is that it is capable of being used with direct currents.[1] Thus the high accuracy and stability of turns on a magnetic core, which are the basis of many high-quality alternating current instruments, is also made available to direct current measurements. Sensitivity to direct current ampere-turns or flux is achieved by modulating the magnetic core with superimposed alternating current ampere-turns, with the result that, when a direct current magnetising force is present, even harmonic components of the modulation are generated. Two identical magnetic cores are normally used so that by suitable interconnection of the windings on each core the odd harmonic components tend to cancel one another and only the even harmonics remain.

Although several modulation and detection techniques are available, the circuit of Fig. 7.1 has some advantages in that the sensing cores themselves, and not an independent network, control the modulation frequency, thereby reducing the effects caused by temperature variation of the core characteristics. In operation the voltage across the combined cores holds one of the transistors in its conducting state and the other in its non-conducting state. When saturation of the cores is achieved these states are exchanged. The result is a square wave of modulation whose frequency is dependent on the magnitude of the supply voltage, the number of turns on the sensing cores, and their magnetic characteristics. A frequency of 700–800 Hz is usually chosen.

Although the major components of the fundamental modulating frequency are eliminated by the winding interconnection, some still remain as sharp pulses at the instant of switching, possibly owing to core mismatch. As direct current ampere-turns are imposed however the peak values of these pulses are unilaterally displaced, and this can be measured with a detector sensitive to peak values.

The magnetic shield protects the modulator cores from the leakage fluxes of the ratio windings and from ambient magnetic fields so that the modulator/detector is sensitive only to those currents which actually link with the cores. It also ensures that modulation currents induced in the ratio windings are constrained by a high inductive impedance.

Two factors affect the overall operation of the modulator/detector. One is

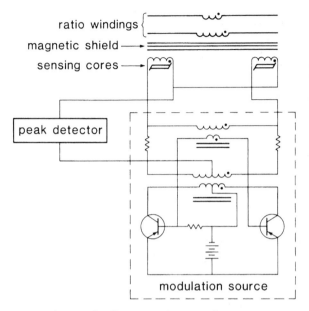

Fig. 7.1 *Ampere-turn detector for direct current comparators*

magnetic noise, which limits the maximum sensitivity to about 10 microampere-turns. The other is the memory effect, which results in a shift in the balance point after a heavy overload. Although the modulation will eventually drive the balance point back to its normal position it is desirable to introduce some means for avoiding this effect, such as by providing a fast responding though not necessarily final automatic balance.

Although inductive coupling is still available to assist in the dynamic balance, it has no effect on the direct current component. This function is undertaken by electronic circuitry.

7.1 Direct current ratio measurements

An important application of the direct current comparator is its use as a current ratio standard for the calibration of resistive shunts and transductors (direct current transformers).[2,3] A typical arrangement is shown in Fig. 7.2. In this circuit a transductor is used to automatically supply about 99% of the secondary current required to bring about ampere-turn balance in the current comparator. The residual imbalance is then sensed by the ampere-turn detector which, through the amplifier, provides the additional current required for balance. The device to be calibrated is connected at points P, S1 and S2. For calibrating a current source device such as a transductor, a small resistor to measure the difference between the two currents would be connected between

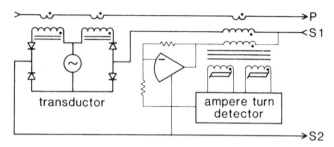

Fig. 7.2 *Direct current ratio standard*

S1 and S2. To calibrate a shunt, a standard resistor would be used to permit a potentiometric measurement. And to reduce the current still further an additional current comparator can be connected in cascade.

A current comparator and its associated transductor with 30·48 cm diameter apertures and rated for primary currents up to 20 000 amperes are shown in Fig. 7.3. They have ratios of 2000, 1750 and 1500 to 1. For the larger currents they are usually cascaded with a 10/1 ratio comparator to provide a secondary current at the 1 ampere level. The ratio accuracy is better than one part in 1 million.

7.2 Direct current resistance bridge

Standard resistors, particularly in the lower values, are usually configured with two sets of terminals, one set being designated as the current terminals and the other set the potential terminals. The resistance is then defined as the voltage

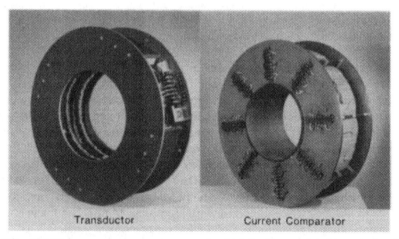

Fig. 7.3 *20 000 ampere direct current comparator (right) with associated transductor (left)*

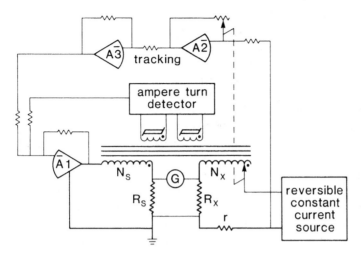

Fig. 7.4 *Direct current comparator resistance bridge*

difference between the potential terminals divided by the current passing through the current terminals. Such resistors can be compared either by measuring the ratio of the two voltages across the resistors when the current is the same, or by measuring the current ratio when the voltages are equal. In the first the greater power dissipation occurs in the larger-valued resistor, while in the other it occurs in the lower-valued. Thus, in general, the first arrangement is preferred for measuring resistors which have values higher than the reference while the latter is more useful for measuring lower-valued resistors.

The most widely known circuit of the first type is the Kelvin double bridge. The measuring circuitry of this bridge however imposes a load on the two resistors being compared, and a secondary balance must be made to correct for the lead resistances. This problem is not present in current comparator bridges where the current circuits are magnetically coupled and no current is present in the potential leads.

The direct current comparator resistance bridge is an embodiment of the second type.[4,5,6] The essential features are shown in Fig. 7.4. Current is supplied from a reversible constant current source to the N_x winding and the unknown resistance R_x. The ampere-turn imbalance is detected and this drives amplifier A1 to supply a current through winding N_s and the reference resistor R_s in such a way as to bring about ampere-turn balance. To relieve the burden on amplifier A1 an additional drive which is proportional to the N_x winding current and the N_x winding turns is provided by the tracking circuit. Comparison of the two resistances is achieved by adjusting the turns N_x so as to null the galvanometer G. At balance

$$R_x = R_s(N_x/N_s)$$

The effect of thermal electromotive forces on the measurement is eliminated by reversing the polarity of the current source and averaging the two results.

The outstanding feature of the bridge is that the first three or four most significant digits of the measurement are determined by winding turns and are thus unaffected by switch contact and other parasitic circuit resistances. Windings N_S and N_X usually have 1000 or 10 000 turns. Additional resolution, up to eight digits overall, is obtained with resistive subdivision of the current by factors of 10, 100, 1000 and 10 000, and an additional winding.

Calibration of the most significant decade of the bridge can be achieved by comparing 11 nominally equal resistors with one another when connected in both the R_S and R_X position and then combining them in various ways to check the various other positions in the decade. A self-check of the internal consistency of the other decades is provided by including an additional negative (-1) position in each decade. This enables the $\times 10$ position of a decade to be compared with a single step of the next more significant decade.

7.3 Direct current comparator potentiometer

The design and construction of accurate, high-resolution direct current potentiometers is limited by the effect that switch contact and other parasitic circuit resistances in the more significant decades may have on the accuracy of the decades of lesser significance.[7] As in the resistance bridge, this problem may be overcome by using turns on a magnetic core for the three or four most significant decades.

The basic configuration of the potentiometer is shown in Fig. 7.5. The output

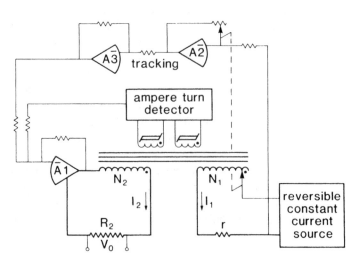

Fig. 7.5   *Direct current comparator potentiometer*

voltage V_0 is given by

$$V_0 = (I_1 N_1)(R_2/N_2)$$

Resistor R_2 and winding turns N_2 are fixed, and current I_1 is adjusted so that the voltage V_0 is given directly by winding turns N_1. Thus to standardise the instrument using a voltage reference such as a standard cell, the winding turns N_1 are set to the standard cell value and the current I_1 is adjusted to make the output voltage equal to the standard cell voltage.

Construction of the potentiometer is similar to that of the resistance bridge. The four decades of least significance are obtained by resistive subdivision and an additional winding as before. Negative (-1) positions are provided in each decade for checking decade-to-decade consistency. In all, some seven decades of useful resolution can be readily provided.

An inherent problem with the current comparator potentiometer is that a small component of the modulation will be present in the output voltage V_0. This can be controlled with more effective magnet shielding and filtering, but it may still be objectionable in some applications.

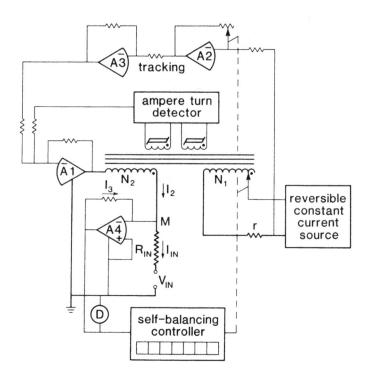

Fig. 7.6 *Direct current comparator differential voltmeter*

7.4 Self-balancing direct current comparator voltmeter

The maximum range of the direct current comparator potentiometer is limited to about 2 volts, and its operation is relatively tedious because the dials must be manually set. To increase the range and provide automatic measurement, a different technique is used.[8]

The basic configuration of the voltmeter is shown in Fig. 7.6. Instead of using a galvanometer to compare voltage with voltage as with a potentiometer, the voltage to be measured V_{IN} is converted into a current by the resistance R_{IN}. This current I_{IN} is compared with the current I_2 from the current comparator winding at node M. Amplifier A4 supplies the difference current I_3 required to hold node M at or close to ground potential. Measurement is made by adjusting winding turns N_1 to bring the current I_3 to zero (and null the detector D). This adjustment is made automatically by the self-balancing controller, which displays the actual turns setting at balance.

If properly calibrated, detector D can be used to indicate any residual imbalance. By providing the means to disable the first, second or third least significant digits of the self-balancing circuitry, detector D can be used to indicate continuously small variations in the quantity being measured.

The input resistance of the instrument is determined by R_{IN}. For I_3 rated at

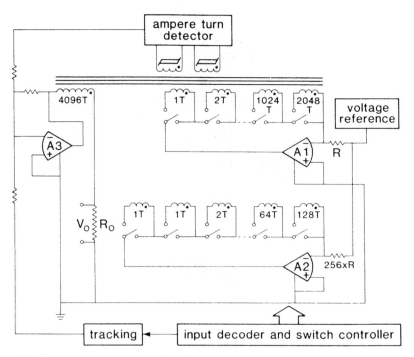

Fig. 7.7   *Direct current comparator digital-to-analogue converter*

l milliampere and R_{IN} equal to 1000 ohms, the input resistance becomes 1000 ohms per volt. R_{IN} is then adjusted accordingly to permit measurement of higher voltages. If a higher input impedance is desired, a voltage follower can be prefaced to the input.

7.5 Current-comparator-based 20-bit digital-to-analogue converter

The basic configuration of a 20-bit digital-to-analogue converter (DAC) which can be used to test the static accuracy of other DACs is shown in Fig. 7.7. Input to the DAC is made by selecting and connecting in series appropriate windings having turns ranging, in powers of 2, from 1 to 2048 to current source amplifier A1, and from 1 to 128 to current source amplifier A2. Amplifier A1 supplies current for the 12 most significant bits and amplifier A2 supplies a current 256 times smaller in magnitude for the 8 remaining bits. An additional one-turn winding completes the binary count to 2^{20}.[9]

The input ampere-turns are balanced by current in a 4096-turn winding supplied by amplifier A3 which in turn is driven by the ampere-turn detector. The voltage output V_0, which is proportional to the digital input, is developed across the resistor R_0 connected in series with the 4096-turn winding.

The input decoder and switch controller converts the decimal input to binary for activating the switches and also provides a tracking voltage to amplifier A3.

References

1 N. L. Kusters, W. J. M. Moore, and P. N. Miljanic. 'A current comparator for precision measurement of D-C ratios.' *Communications and Electronics*, 70, January 1964, pp. 22 7

2 M. P. MacMartin and N. L. Kusters. 'A self-balancing direct current comparator for 20 000 amperes.' *IEEE Trans. Magnetics*, **MAG-1**(4), December 1965, pp. 396 402

3 N. L. Kusters and M. P. MacMartin. 'A direct-current-comparator bridge for measuring shunts up to 20 000 amperes.' *IEEE Trans. Instrumentation and Measurement*, **IM-18**(4), December 1969, pp. 226 71

4 M. P. MacMartin and N. L. Kusters. 'A direct-current-comparator ratio bridge for four-terminal resistance measurements.' *IEEE Trans. Instrumentation and Measurement*, **IM-15**(4), December 1966, pp. 212 20

5 N. L. Kusters and M. P. MacMartin. 'Direct-current comparator bridge for resistance thermometry.' *IEEE Trans. Instrumentation and Measurement*, **IM-19**(4), November 1970, pp. 291 7

6 N. L. Kusters and M. P. MacMartin. 'A direct-current-comparator bridge for high resistance measurement.' *IEEE Trans. Instrumentation and Measurement*, **IM-22**(4), December 1973, pp. 382 6

7 M. P. MacMartin and N. L. Kusters. 'The application of the direct current comparator to a seven-decade potentiometer.' *IEEE Trans. Instrumentation and Measurement*, **IM-17**(4), December 1968, pp. 263 8

8 N. L. Kusters and M. P. MacMartin. 'A self-balancing digital differential voltmeter based on the direct-current comparator.' *IEEE Trans. Instrumentation and Measurement*, **IM-24**(4), December 1975, pp. 331 5

9 E. So. 'A current-comparator-based 20-bit digital-to-analog converter.' *IEEE Trans. Instrumentation and Measurement*, **IM-34**(2), June 1985, pp. 278 82

Index

.